SolidWorks 2018 Black Book

By
Gaurav Verma
Matt Weber
(CADCAMCAE Works)

Edited by
Kristen

ISBN # 978-1-988722-21-4

DEDICATION

To teachers, who make it possible to disseminate knowledge
to enlighten the young and curious minds
of our future generations

To students, who are the future of the world

THANKS

To my friends and colleagues

To my family for their love and support

Training and Consultant Services

At CADCAMCAE WORKS, we provides effective and affordable one to one online training on various software packages in Computer Aided Design(CAD), Computer Aided Manufacturing(CAM), Computer Aided Engineering (CAE), Computer programming languages(C/C++, Java, .NET, Android, Javascript, HTML and so on). The training is delivered through remote access to your system and voice chat via Internet at any time, any place, and at any pace to individuals, groups, students of colleges/universities, and CAD/CAM/CAE training centers. The main features of this program are:

Training as per your need

Highly experienced Engineers and Technician conduct the classes on the software applications used in the industries. The methodology adopted to teach the software is totally practical based, so that the learner can adapt to the design and development industries in almost no time. The efforts are to make the training process cost effective and time saving while you have the comfort of your time and place, thereby relieving you from the hassles of traveling to training centers or rearranging your time table.

Software Packages on which we provide basic and advanced training are:

CAD/CAM/CAE: CATIA, Creo Parametric, Creo Direct, SolidWorks, Autodesk Inventor, Solid Edge, UG NX, AutoCAD, AutoCAD LT, EdgeCAM, MasterCAM, SolidCAM, DelCAM, BOBCAM, UG NX Manufacturing, UG Mold Wizard, UG Progressive Die, UG Die Design, SolidWorks Mold, Creo Manufacturing, Creo Expert Machinist, NX Nastran, Hypermesh, SolidWorks Simulation, Autodesk Simulation Mechanical, Creo Simulate, Gambit, ANSYS and many others.

Computer Programming Languages: C++, VB.NET, HTML, Android, Javascript and so on.

Game Designing: Unity.

Civil Engineering: AutoCAD MEP, Revit Structure, Revit Architecture, AutoCAD Map 3D and so on.

We also provide consultant services for Design and development on the above mentioned software packages

For more information you can mail us at:
cadcamcaeworks@gmail.com

Table of Contents

Chapter 3 : Advanced Dimensioning and Practice

Chapter 4 : 3D Sketch and SolidModeling

Chapter 5 : Solid Editing and Practical

Chapter 6 : Advanced Solid Modeling New

Chapter 7 : Assembly and Motion Study

Chapter 8 : Advanced Assembly, Practical and Practice

Chapter 9 : Surfacing and Practice

Chapter 10 : Drawing and Practice

Chapter 11 : AnalysisXpress

Chapter 12 : Mold Tools

Preface

SolidWorks 2018 is a parametric, feature-based solid modeling tool that not only unites the three-dimensional (3D) parametric features with two-dimensional (2D) tools, but also addresses every design-through-manufacturing process. The continuous enhancements in the software has made it a complete PLM software. The software is capable of performing analysis with an ease. Its compatibility with CAM software is remarkable. Based mainly on the user feedback, this solid modeling tool is remarkably user-friendly and it allows you to be productive from day one.

The **SolidWorks 2018 Black Book** is the 5th edition of our series on SolidWorks. With lots of additions and thorough review, we present the book to help professionals as well as learners in creating some of the most complex solid models. The book follows a step by step methodology. In this book, we have tried to give real-world examples with real challenges in designing. We have tried to reduce the gap between university use of SolidWorks and industrial use of SolidWorks. In this edition of book, we have included many new features of SolidWorks like 3D printing, SolidWorks MBD (Model Based Definition), Magnetic Mates, Asset Builder, 3D PDFs, SolidWorks CAM and so on. There are two new chapters added in this book which were not in the previous edition. The book covers almost all the information required by a learner to master the SolidWorks. The book starts with sketching and ends at advanced topics like Mold Design, Sheetmetal, Weldment, and SolidWorks MBD, SolidWorks CAM. Some of the salient features of this book are :

In-Depth explanation of concepts

Every new topic of this book starts with the explanation of the basic concepts. In this way, the user becomes capable of relating the things with real world.

Topics Covered

Every chapter starts with a list of topics being covered in that chapter. In this way, the user can easy find the topic of his/her interest easily.

Instruction through illustration

The instructions to perform any action are provided by maximum number of illustrations so that the user can perform the actions discussed in the book easily and effectively. There are about 1090 illustrations that make the learning process effective.

Tutorial point of view

At the end of concept's explanation, the tutorial make the understanding of users firm and long lasting. Almost each chapter of the book has tutorials that are real world projects. Moreover most of the tools in this book are discussed in the form of tutorials.

Project

Free projects and exercises are provided to students for practicing.

For Faculty

If you are a faculty member, then you can ask for video tutorials on any of the topic, exercise, tutorial, or concept.

New Addition

If anything is added in this edition but is not available in the previous editions, then it is displayed with symbol **New** in table of content.

Formatting Conventions Used in the Text

All the key terms like name of button, tool, drop-down etc. are kept bold.

Free Resources

Link to the resources used in this book are provided to the users via email. To get the resources, mail us at *cadcamcaeworks@gmail.com* with your contact information. With your contact record with us, you will be provided latest updates and informations regarding various technologies. The format to write us mail for resources is as follows:

Subject of E-mail as *Application for resources of   book*.
Also, given your information like
Name:
Course pursuing/Profession:
Contact Address:
E-mail ID:

Note: We respect your privacy and value it. If you do not want to give your personal informations then you can ask for resources without giving your information.

About Authors

The author of this book, Matt Weber, has written many books on CAD/CAM/CAE available already in market. **SolidWorks Simulation Black Books** are one of the most selling books in SolidWorks Simulation field. The author has hands on experience on almost all the CAD/CAM/CAE packages. Besides that he is a good person in his real life, helping nature for everyone. If you have any query/doubt in any CAD/CAM/CAE package, then you can contact the author by writing at cadcamcaeworks@gmail.com

The author of this book, Gaurav Verma, has written and assisted in more than 14 titles in CAD/CAM/CAE which are already available in market. He has authored **AutoCAD Electrical Black Books** which are available in both **English** and **Russian** language. He has provided consultant services to many industries in US, Greece, Canada, and UK.

For Any query or suggestion

If you have any query or suggestion, please let us know by mailing us on *cadcamcaeworks@gmail.com*. Your valuable constructive suggestions will be incorporated in our books and your name will be addressed in special thanks area of our books on your confirmation.

Chapter 1

Starting with SolidWorks

Topics Covered

The major topics covered in this chapter are:

- *Installing SolidWorks 2018.*
- *Starting SolidWorks 2018.*
- *Starting a new document.*
- *Terminology used in SolidWorks.*
- *Opening a document.*
- *Closing documents.*
- *Basic Settings for SolidWorks*
- *Workflow in Industries using the SolidWorks*

INSTALLING SOLIDWORKS 2018

- If you are installing SolidWorks using the CD/DVD provided by Dassault Systemes then go to the folder containing **setup.exe** file and then right click on **setup.exe** in the folder. A shortcut menu is displayed on the screen; refer to Figure-1.

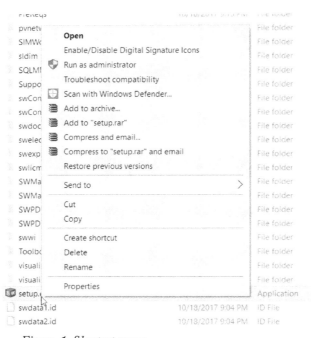

Figure-1. Shortcut menu

- Select the **Run as Administrator** option from the menu displayed; refer to Figure 1.
- Select the **Yes** button from the dialog box displayed. The **SolidWorks 2018 Installation Manager** will be displayed. Follow the instructions given in the dialog box. Note that you must have the **Serial Number** with you to install the application. To get more about installation, double-click on the **Read Me** documentation file in the Setup folder.
- If you have downloaded the software from Internet, then you are required to browse in the **SolidWorks Download** folder in the **Documents** folder of Computer. Open the folder of latest version of software and then run **setup.exe**. Rest of the procedure is same.

STARTING SOLIDWORKS 2018

- To start SolidWorks in Windows 10 from **Start** menu, click on the **Start** button in the Taskbar at the bottom left corner, click on the **SolidWorks 2018** folder. In this folder, select the SolidWorks 2018 icon; refer to Figure-2.

Figure-2. Start menu

- While installing the software, if you have selected the check box to create a desktop icon, then you can double-click on that icon to run the software.
- If you have not selected the check box to create the desktop icon but want to create the icon on desktop, then drag and drop the **SolidWorks 2018** icon from the Start menu on the desktop.

After you perform the above steps, the SolidWorks 2018 application window will be displayed; refer to Figure-3.

Figure-3. SolidWorks 2018 application window

STARTING A NEW DOCUMENT

You can start SolidWorks by four ways:

1. Click on the **Part** button in the **Welcome** dialog box; refer to Figure-3.
Or
2. Click on the **New** button in the **Menu Bar** .
Or
3. Move the cursor on the left arrow near the **SolidWorks icon**; refer to Figure-4 and then click on the **File** menu button and click on the **New** button; refer to Figure-4.
Or
4. Press **CTRL** and **N** together from the Keyboard.

Figure-4. File menu

- On performing any one of the above steps, the **New SolidWorks Document** dialog box will be displayed as shown in Figure-5.

Figure-5. New SOLIDWORKS document

- If you are creating first document after installing SolidWorks then **Units and Dimension Standard** dialog box will be displayed; refer to Figure-6.

Figure-6. Units and Dimension Standard dialog box

- Select the unit system and dimension standards that you want to use while creating documents in SolidWorks. These options will be set as default. You can change these parameters later. Click on the **OK** button from the dialog box.
- There are three buttons available in this dialog box; **Part, Assembly** and **Drawing**.

The **Part** button is used to create Solid or Surface models.

The **Assembly** button is used to create Assemblies.

The **Drawing** button is used to create drawings from the solid/surface models or assemblies.

You will learn more about solids, surfaces, assemblies, and drawings later in the book.

Note that the building blocks of CAD are solid models. In SolidWorks, solid models are created by using the tools available in the **Part** mode. You can start with the **Part** mode by selecting the **Part** button in the **New SOLIDWORKS Document** dialog box.

- Double-click on the **Part** button to start the Solid/surface modeling environment of SolidWorks. On doing so, the application interface will be displayed as shown in Figure-7.

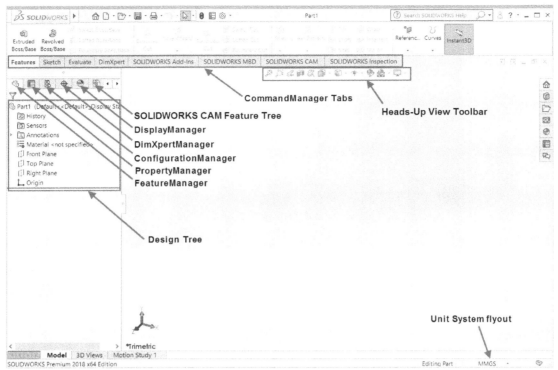

Figure-7. Application interface

The tools available in the **Part, Assembly** and **Drawing** mode are compiled in the form of **CommandManagers**. To display or hide a **CommandManager** from **Ribbon**, right-click on a **CommandManager** in the **Ribbon**. A shortcut menu will be displayed as shown in Figure-8. Select the **CommandManager** that you want to add or remove from the **Ribbon**. Various **CommandManagers** available in SolidWorks are discussed next.

Figure-8. Shortcut menu for CommandManager

Part Mode CommandManagers

A number of **CommandManagers** can be invoked in the **Part** mode. These **CommandManagers** with their functioning are described next.

Sketch CommandManager

The tools available in this **CommandManager** are used to draw sketches for creating solid/surface models. This **CommandManager** is also used to add relations and smart dimensions to the sketched entities. The **Sketch CommandManager** is shown in Figure-9.

Figure-9. Sketch CommandManager

Features CommandManager

This **CommandManager** provides all modeling tools that are used for feature-based solid modeling. The **Features CommandManager** is shown in Figure-10.

Figure-10. Features CommandManager

DimXpert CommandManager

This **CommandManager** is used to add dimensions and tolerances to the features of a part. The **DimXpert CommandManager** is shown in Figure-11.

Figure-11. DimXpert CommandManager

Sheet Metal CommandManager

The tools in this **CommandManager** are used to create the sheet metal parts. The **Sheet Metal CommandManager** shown in Figure-12. If this **CommandManager** is not added in the **Ribbon**, then right-click on any of the CommandManager tab and select the **Sheet Metal** option from the menu; refer to Figure-13.

Figure-12. Sheet Metal CommandManager

Figure-13. Adding hidden tabs in Ribbon

Mold Tools CommandManager

The tools in this **CommandManager** are used to design a mold and split core and cavity steel. The **Mold Tools CommandManager** is shown in Figure-14.

Figure-14. Mold Tools CommandManager

Evaluate CommandManager

This **CommandManager** is used to measure entities, perform analysis and so on. The **Evaluate CommandManager** is shown in Figure-15.

Figure-15. Evaluate CommandManager

Surfaces CommandManager

This **CommandManager** is used to create complicated surface features. The **Surfaces CommandManager** is shown in Figure-16.

Figure-16. Surfaces CommandManager

Direct Editing CommandManager

This **CommandManager** consists of tools (Figure-17) that are used for editing a feature.

Figure-17. Direct Editing CommandManager

Data Migration CommandManager

This **CommandManager** consist of tools (Figure-18) that are used to work with the models created in other packages or in different environments.

Figure-18. Data Migration CommandManager

Weldments CommandManager

This **CommandManager** is used to create welding joints in the model and assembly. The **Weldments CommandManager** is shown in Figure-19.

Figure-19. Weldments CommandManager

SOLIDWORKS MBD CommandManager

This **CommandManager** is used to apply Model Based Dimension which means the dimensions are directly applied to model while skipping the steps of generating drawings. Use SolidWorks MBD in manufacturing industry requires electronic gadgets at shop floor for production and quality checks. These gadgets should be capable of displaying CAD eDrawings. The CommandManager is shown in Figure-20.

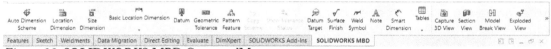

Figure-20. SOLIDWORKS MBD CommandManager

Render Tools CommandManager

The tools in **Render Tools CommandManager** are used to render the image of current model using appearance parameters specified. The CommandManager is shown in Figure-21.

Figure-21. Render Tools CommandManager

SOLIDWORKS CAM CommandManager

The tools in this CommandManager are used to setup NC programming model for the part and generate the codes for CNC machines. The **SOLIDWORKS CAM CommandManager** is shown in Figure-22.

Figure-22. SOLIDWORKS CAM CommandManager

Assembly Mode CommandManagers

The tools in **CommandManagers** of the **Assembly** mode are used to assemble the components. The **CommandManagers** in the **Assembly** mode are discussed next.

Assembly CommandManager

This **CommandManager** is used to insert a component and apply various types of mates to the assembly. The **Assembly CommandManager** is shown in Figure-23.

Figure-23. Assembly CommandManager

Layout CommandManager

The tools in this **CommandManager** (Figure-24) are used to create and edit blocks.

Figure-24. Layout CommandManager

Drawing Mode CommandManagers

You can invoke a number of **CommandManagers** in the **Drawing** mode. The **CommandManagers** that are extensively used during the designing process in this mode are discussed next.

View Layout CommandManager

This **CommandManager** is used to generate the drawing views of an existing model or an assembly. The **View Layout CommandManager** is shown in Figure-25.

Figure-25. View Layout CommandManager

Annotation CommandManager

The **Annotation CommandManager** is used to generate the model items and to add notes, balloons, geometric tolerance, surface finish symbols, and so on to the drawing views. The **Annotation CommandManager** is shown in Figure-26.

Figure-26. Annotation CommandManager

The commands available in these CommandManagers will be discussed one by one later in this book.

OPENING A DOCUMENT

Like creating new documents, there are many ways to open documents. Some of the ways are discussed next.

1. Click on the **Open a Document** link in the Task Pane; refer to Figure-3.
Or
2. Click on the **Open** button in the **Menu Bar** 📄 .
Or
3. Move the cursor on the **SolidWorks icon**; refer to Figure-4 and then click on the **File > Open** button from the menu; refer to Figure-4.
Or
4. Press **CTRL** and **O** together from the Keyboard.
• After performing any of the above step, the **Open** dialog box will be displayed; refer to Figure-27.

Figure-27. Open dialog box

• Select the file type of your file from the **File Type** fly-out `Custom (*.prt;*.asm;*.drw;*.sld|` ▼ in the bottom right corner of the dialog box.

- Browse to the folder in which you have saved the file and then double-click on it to open.

Most of the time people close the application to close the current file and then restart the application to start working again. In SolidWorks, you can close the current file while the application remains open. The steps to do so are given next.

CLOSING A DOCUMENT

To close a document, there are two ways displayed in Figure-28.

- Open the **File** menu and then click on the **Close** option from it.

Or

- Click on the **Close** button ✕ at the top-right of the current viewport. (Viewport is the area in which the model is displayed.)

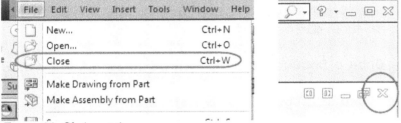

Figure-28. Closing options

- If you have done some changes in the document then the dialog box will be displayed, prompting you to save the document; refer to Figure-29.

Figure-29. Save prompt

- Click on the **Save all** button to save the changes or click on the **Don't Save** button to reject the changes.

Till this point you know, starting SolidWorks; Creating, Opening, and Closing documents; and you have idea about the interface of SolidWorks. Now, we will discuss about some basic settings that are required for easy working with SolidWorks.

BASIC SETTINGS OF SOLIDWORKS

All the settings of SolidWorks are compiled in the **Options** dialog box. The steps to change the settings for SolidWorks are given next.

- Click on the **Options** option in the **Tools** menu or click on the **Options** button ⚙ from the Menu Bar.
- On performing the above step, the **System Options** dialog box will be displayed as shown in Figure-30. Note that if you have a document opened then the **Document Properties** tab is also added with the **System Options** tab. To get the detail about each and every option, you need to refer to SolidWorks Help Documentation. In this section, we will discuss about some of the important options that are generally required.

Figure-30. Partial System Options dialog box

- Click on the **Sketch** option in the left of the dialog box, select the **Enable on screen numeric input on entity creation** check box from the right to enter dimensions while creating the sketch. Also, select the **Create dimension only when value is entered** check box to create dimensions only when you have manually entered the dimension value.
- If you want to use only fully defined sketches for creating features in SolidWorks then select the **Use Fully defined sketches** check box. Fully defined sketches are those sketch which have all their entities dimensioned or constrained.

- Click on **Relations/Snaps** in the left of the dialog box and then select the **Enable snapping** check box to enable auto snapping to the key points.
- Click on the **Documents Properties** tab if you have any document opened in the viewport. Click on the **Units** option from the left. The **Options** dialog box will be displayed as shown in Figure-31.
- Select the desired radio button from the right to set the unit system for the current document.
- Click on the **Drafting Standard** option from the left area of the dialog box and select the desired dimensioning standard from the **Overall drafting standard** drop-down; refer to Figure-32.

Figure-31. Document properties tab

Figure-32. Overall drafting standard drop-down

- Similarly, you can set the other options in the dialog box. Note that we will be revisiting this dialog box many times in this book.
- Click on the **OK** button from the bottom of the dialog box to save the settings.

You can also change the units of document by selecting the desired option from the list displayed on clicking on the **Unit system** flyout at the bottom of the viewport; refer to Figure-33.

Figure-33. Unit system flyout

MOUSE BUTTON FUNCTION

Rotate View (parts and assemblies only)
- To rotate the model view: Drag with the middle mouse button in desired direction.
- To rotate about a vertex, edge, or face: Middle-click a vertex, edge, or face; then middle-drag the pointer.

Pan
Hold down Ctrl and drag with the middle mouse button. (In an active drawing, you do not need to hold down Ctrl.)

Zoom In/Out
Hold down Shift and drag with the middle mouse button. Or scroll the mouse wheel forward and backward on Wheel-type Mouse.

LOADING ADD-INS

Add-Ins are used to allow external functions in SolidWorks. Like, you can use SolidWorks Electrical, SolidWorks Simulation, SolidWorks PCB etc. The procedure to load Add-Ins is given next.

- Click on the **Add-Ins** option from the **Options** drop-down in the Menu Bar; refer to Figure-34. The **Add-Ins box** will be displayed; refer to Figure-35.

Figure-34. Add-Ins option

Figure-35. Add-Ins box

- Select the left side check box for Add-In that you want to load now. If you want to load an Add-In at startup of SolidWorks then select the right-side check box for it.
- Click on the **OK** button from dialog box to apply the settings.

SEARCH TOOLS

The **Search box** at the top-right corner of the application window is a multipurpose tool. You can use this search box to search for help content, commands, cad models, training files etc. The method to use **Search box** for searching commands is given next. You can apply the same method to search other things.

- Hover the cursor on **SOLIDWORKS** logo at the top-left corner of application window and click on the **Commands** option from the **Search** cascading menu of **Help** menu; refer to Figure-36.

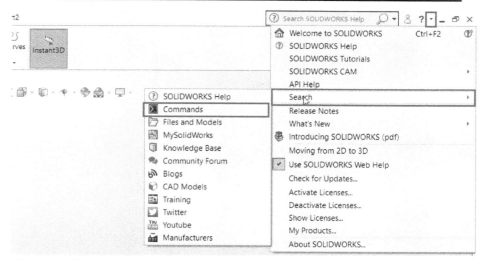

Figure-36. Commands search option

- Click in the **Search box** and type few characters of command that you are searching. A list of tools with typed characters will be displayed; refer to Figure-37.

Figure-37. List of tools in search box

- Click on the eye icon displayed next to tool in the list. The cursor will move to the tool location and an arrow will be displayed pointing to the tool; refer to Figure-38. If you click on the tool in the list then it will be activated directly.

Figure-38. Location of tool displayed

WORKFLOW IN SOLIDWORKS

The first step in SolidWorks is to create a sketch. After creating sketch of the desired feature, we create solid or surface model from that sketch. After doing the desired operations on the solid/surface model, we go for assembly or analyses. After, we are satisfied with the assembly/ analyses, we create the engineering drawings from the model to allow manufacturing of the model into a real world object.

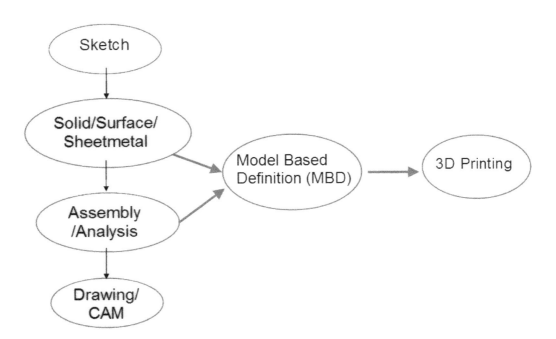

SELF ASSESSMENT

Q1. If you have downloaded the SolidWorks Setup files from Internet then the files will be available in Downloads folder of Windows by default. (T/F)

Q2. We cannot create the desktop icon of SolidWorks if we have not opted for it while installing. (T/F)

Q3. Status Bar in SolidWorks window also displays the tips related to current tool. (T/F)

Q4. Which of the following operation results in display of **New SolidWorks Document** dialog box?
(a)   Press CTRL+N
(b)   Click on **New** button from Menu Bar
(c)   Click on **New Document** link button from Task Pane.
(d)   All of the above

Q5. Which of the following is not an option in the New SolidWorks Document dialog box?
(a)   Part
(b)   Assembly
(c)   Drawing
(d)   Sketch

Q6. The .... CommandManager provides all modeling tools that are used for feature-based solid modeling.

Q7. The .... CommandManager is used to add dimensions and tolerances to the features of a part.

Q8. The .... CommandManager is used to insert a component and apply various types of mates to the assembly.

Q9. Write down the steps to close the current document in SolidWorks.

Q10. How can we change the unit system of current SolidWorks document?

Q11. What is the purpose of Add-Ins in SolidWorks?

Q12. The .... check box forces SolidWorks to use only fully defined sketches for creating features.

Q13. The ....CommandManager is used to create codes for CNC machines based on part model.

Q14. What is the purpose of **Welcome** dialog box?

FOR STUDENT NOTES

FOR STUDENT NOTES

Chapter 2

Sketching
Basic to Advanced

Topics Covered

The major topics covered in this chapter are:

- *Basics for Sketching.*
- *Entity creation tools.*
- *Entity Editing tools.*
- *Dimensioning and Constraining.*
- *3D Sketching.*
- *Printing and exporting sketch.*

BASICS FOR SKETCHING

In Engineering, sketching do not mean sketches of birds or animals. It means sketches that are based on real dimensions of real-world objects. Now, we are going to work with geometric entities like; line, circle, arc, ellipse, and so on which you have used in your school and engineering. But this time, we will be using the software in place of pencil, ruler and other geometry tools. Note that the sketching is the base for most of 3D Models so you should be proficient in sketching.

To start with sketching, we must have a good understanding of planes in SolidWorks. Next figure (Figure-1) shows the names of planes and their respective faces in Isometric View.

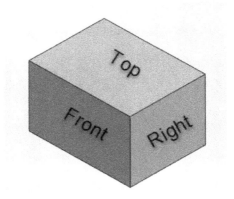

Figure-1. Planes

In SolidWorks, the planes are displayed in the same orientation as shown in the above figure. To check the planes of SolidWorks, click on the planes (Front Plane, Top Plane, Right Plane) in the **FeatureManager Design Tree** after starting Part modeling, refer to Figure-2.

Figure-2. Planes in Property-Manager

- To show these planes, select them one by one while holding the CTRL key and right-click. A shortcut menu will be displayed; refer to Figure-3.

Figure-3. Shortcut menu on right clicking on planes

- Select the **Show/Hide** button 👁 from the shortcut menu; the planes will be displayed. To hide planes, click on the **Show/Hide** button again.

SKETCHING PLANE

In a CAD software, everything is referenced to other entity like a line created must be referenced to any other geometry so that you can clearly define the position of line with respect to the other geometry. But, what if there is no geometry in the sketch to reference from. In these cases, we have tools to create reference geometries like reference planes, axes, points, curves etc. Out of these reference geometries, the sketching plane acts as foundation for other geometries. By default, there are three plane perpendicular to each other in SolidWorks; refer to Figure-4.

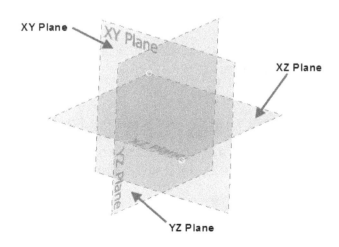

Figure-4. Sketching planes in SolidWorks

Relation between sketch, plane, and 3D model

Sketch has a direct relationship with planes and the outcome which is generally a 3D model. Check the Figure-5. In this figure, rectangle is created on the XY Plane which is also called **Front** plane. A circle is created on the YZ Plane which is also called **Right** plane. A polygon is created on the XZ plane which is also called **Top** plane. In a 3D model, the geometry seen from the Front view should be drawn on the **Front** plane. Similarly, geometry seen from the Right view should be drawn at **Right** plane and geometry seen from the Top view should be drawn at the **Top** plane.

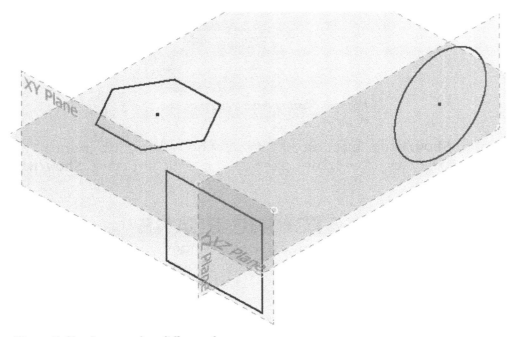

Figure-5. Sketches created on different planes

We will learn more about planes at the beginning of 3D Modeling. We will now start with sketching tools of SolidWorks.

Now, we are ready to start with sketch.

Start SolidWorks, create a new Part document and then click on the **Sketch** tab in the **CommandManagers**. The tools for sketching are displayed in the **Sketch CommandManager**.

STARTING SKETCH

- Click on the **Sketch** button at the left in the **Sketch** CommandManager. Three main planes are displayed with their name in the viewport; refer to Figure-6.

Figure-6. Sketching on default planes

- Click on the desired plane from the viewport or click on the arrow displayed in the top-left corner of the viewport and select the name of the plane from the list of features; refer to Figure-33.
- On clicking a plane, the selected plane will become parallel to the screen. Now, we are ready to draw sketch on the plane. The **Sketch CommandManager** is divided into sections depending on the functions of the tools. Figure-7 shows the **Sketch CommandManager** and the division of tools.

Figure-7. Sketch CommandManager

We will start with the sketch creation tools and then one by one, we will discuss other tools. But, before that it is important to understand the tools in Heads-Up View Toolbar.

HEADS-UP VIEW TOOLBAR

The **Heads-up View Toolbar** contains tools to change the view and orient the modeling area; refer to Figure-8.

Figure-8. Heads-up View toolbar

The functions of the tools displayed in above figure are discussed next.

Zoom to Fit : This tool is used to display all the objects created in the viewport. To use this tool, click on the tool once. The objects will automatically fit in the current viewport.

Zoom to Area : This tool is used to display a specific area in the viewport zoomed to the full extent. To use this tool, click on it. The cursor will change to a zoom box selection cursor and you are asked to create a boundary box surrounding the entities you want to zoom in. Click to specify the starting point of the zoom box and then drag to the point till where you want to complete the zoom box. The area in the box will zoom automatically. Figure-9 shows a zoom box drawn to zoom.

Figure-9. Zoom box

Previous View : This tool is used to zoom to the previous level. To use this tool, click on it once. The viewport will be displayed at the previous level.

Section View : This tool is used to display section of a solid model. (Section is created when you cut a solid from a plane. It is mainly used to see the inside of the model.)

Dynamic Annotation View : This tool is used to display or hide the annotations applied to the model in Part/Assembly environment. You will learn more about this tool in Solid Modeling.

View Orientation : This tool is used to change the view orientation of the model. When you click on this button, a toolbox will be displayed as shown in Figure-10. The buttons perform the action mentioned in the figure.

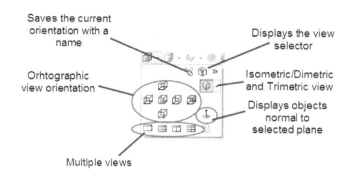

Figure-10. View orientation toolbox

Display Style : These tools are used to display the model is different styles, like shaded, hidden, no hidden and so on.

Hide/Show Items : When you click on this button, a tool box will be displayed with various toggle buttons. These buttons allow to display or hide key feature like, geometric relations, center lines, annotations, grids, and so on. To enable or disable the view of a key feature, select the respective button.

Next three buttons will be discussed later in the book.

SKETCH CREATION TOOLS

The standard tools to draw sketch entities are available in this section of **Ribbon**. These tools are discussed next.

Line Tools

There are three tools in the **Line** drop-down to create different type of lines; **Line**, **Centerline** and **Midpoint Line**; refer to Figure-11.

Figure-11. Line drop-down

Line

We use this tool to create every type of lines required in creating sketch. The procedure to create line is explained in the next steps:

- Click on the **Line** tool from the drop-down. The **Insert Line PropertyManager** will be displayed; refer to Figure-12.

Figure-12. Insert Line PropertyManager

- The radio buttons in the **Orientation** rollout are used to set the orientation of the line before drawing it. There are four radio buttons available in this rollout; **As sketched**, **Horizontal**, **Vertical**, and **Angle**. The **As sketched** radio button is selected by default. So, you do not need to define the orientation of line, you can just start creating line. Click in the drawing area to create the line.

Select the **Horizontal** radio button, if you want to create horizontal line. On selecting this radio button, the **Parameters** rollout will be displayed at the bottom of the rollout. Specify the length of the line in the edit box displayed in the **Parameters** rollout and click to specify start point of line to create the line with specified length.

Select the **Vertical** radio button, if you want to create the vertical line. This works in the same way as the horizontal option works.

Select the **Angle** radio button, if you want to create the lines at the specified angle. On selecting this radio button, two edit boxes will become available. Specify the length of the line and angle of the line in the respective edit boxes; refer to Figure-13.

Figure-13. Parameters rollout

- The check boxes in the **Options** rollout are used to modify the properties of line while creating it. On selecting the **For construction** check box, you will create a line in construction mode. On selecting the **Infinite length** check box, you can create a line of infinite length. The **Midpoint line** check box was added in SolidWorks 2015. On selecting this check box, you can create a line with the help of mid point and end point. The **Add dimensions** check box is used to add dimensions while creating the line. Note that the **Add dimensions** check box is displayed only when you have selected the **Enable on screen numeric input on entity creation** check box from the **Options** dialog box as discussed in previous chapter. You can use the **Add dimension** check box only when **Angle** radio button is selected along with Infinite length check box.
- Most of the time, we use the **As sketched** radio button to create lines in sketch. Select the **As sketched** radio button.
- When you move the cursor in the viewport. By default the cursor snaps to the key points like horizontal/ vertical to coordinate system, coincident to the coordinate system and so on. Click to specify the start point on the screen. Figure-14 shows the creation of line.

Figure-14. Line creation

- Press ESC button from the keyboard to exit the tool.

Now, stop reading the book and first practice on the tool by using all the options one by one. From now onwards, you should practice on the tool in every possible manner after it has been discussed because we are not in theory business!!

Centerline

The **Centerline** tool is used to create center line in the viewport. The procedure to create a center line is the same as line creation. Click on this tool, specify the start point of the center line and then specify the end point of the center line.

Midpoint Line

The **Midpoint Line** tool was added in SolidWorks 2015. Generally, we draw a line in SolidWorks by using the start point and end point but we can also draw a line by using the mid point and end point of the line. The method to use this tool is given next.

- Click on the **Midpoint Line** tool from the **Line** drop-down. The **Insert Line PropertyManager** will be displayed as shown in Figure-15.

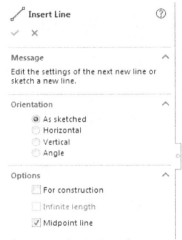

Figure-15. Insert Line Property Manager

- Click to specify the mid point of the line. The end point of line will get attached to the cursor.
- Move the cursor in desired direction and click to specify the end point of line. You can also enter the value in the edit box displayed with the cursor while specifying end point; refer to Figure-16.

Figure-16. Midpoint line creation

Rectangle

There are five tools in the **Rectangles** drop-down; **Corner Rectangle**, **Center Rectangle**, **3 Point Corner Rectangle**, **3 Point Center Rectangle**, and **Parallelogram**; refer to Figure-17.

Figure-17. Rectangles drop-down

The procedures to use these tools are discussed next.

Corner Rectangle

- Click on the **Corner Rectangle** tool from the drop-down. The **Rectangle PropertyManager** will be displayed as shown in Figure-18.

Figure-18. Rectangle PropertyManager

- Click in the viewport to specify the first point. The current dimensions will be displayed along the lines of rectangle. Specify the desired value and press the **Tab** key from the Keyboard to switch to the other dimension.
- Specify the other dimension and then press **Enter** from the Keyboard. The rectangle will be created with the specified dimensions.
- Select the **Add dimensions** check box if you want to apply the dimensions while creating the rectangle.
- You can add the construction lines for rectangle being created by selecting the **Add construction lines** check box. There are two ways by which you can add the construction lines in a rectangle; refer to Figure-19.

Figure-19. Rectangle with construction lines

- You can switch to the other types of rectangle by using the five buttons available in the top section of the **PropertyManager**.
- Note that from SolidWorks 2017 onwards, the closed loop sketches are displayed as shaded like in Figure-20.

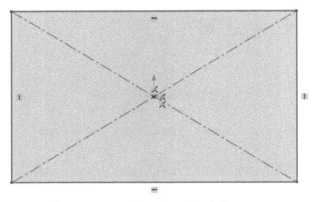

Figure-20. Closed loop sketch displayed shaded

- If you do not want the closed loops to be displayed shaded, then click on the **Shaded Sketch Contours** toggle button from the **Sketch** tab in the **Ribbon** to de-select the button; refer to Figure-21.

Figure-21. Shaded Sketch Contour button

Center Rectangle

- Click on the **Center Rectangle** tool from the drop-down. The **Rectangle PropertyManager** will be displayed similar to the one displayed earlier. In this **PropertyManager**, the **Center Rectangle** button is selected by default.
- Click to specify the center point of the rectangle.
- Specify the corner point of the rectangle. You can specify any of the corner point by moving the cursor in desired direction.

3 Point Corner Rectangle

- Click on the **3 Point Corner Rectangle** tool from the drop-down. The **Rectangle PropertyManager** will be displayed similar to the one displayed earlier. In this **PropertyManager,** the **3 Point Corner Rectangle** button is selected by default.
- Click to specify the starting point of the rectangle.
- Click to specify the end point of the base line.
- Click to specify the end point for the height. Refer to Figure-22 for procedure.

Figure-22. 3Point Rectangle Creation

3 Point Center Rectangle

- Click on the **3 Point Center Rectangle** tool from the drop-down. The **Rectangle PropertyManager** will be displayed similar to the one displayed earlier. In this **PropertyManager,** the **3 Point Center Rectangle** button is selected by default.
- Click to specify the center point of the rectangle.
- Click to specify the half length of the base line
- Click to specify the half length of vertical line. Refer to Figure-23 for the procedure.

Figure-23. 3Point Center Rectangle Creation

Parallelogram

- Click on the **Parallelogram** tool from the drop-down. The **Rectangle PropertyManager** will be displayed similar to the one displayed earlier. In this **PropertyManager,** the **Parallelogram** button is selected by default.
- Click to specify the start point of the base line.
- Click to specify the end point of the base line.
- Click to specify the end point of the line defining height of the parallelogram. While defining this line, you can move the cursor in left/right and vertical direction.

Slot

There are four tools in the **Slot** drop-down; **Straight Slot, Center Straight Slot, 3 Point Arc Slot**, and **Centerpoint Arc Slot**; refer to Figure-24.

Figure-24. Slots drop-down

The tools in this drop-down are explained next.

Straight Slot

• Click on the **Straight Slot** tool from the drop-down. The **Slot PropertyManager** will be displayed as shown in Figure-25.

Figure-25. Slot PropertyManager

• Click on the **Center to Center** button or **Overall Length** button to set the length dimension style for slot. By default, the **Center to Center** button is selected in **PropertyManager**.
• Click in the viewport to specify the center of the first semi-circle of slot.
• Click to specify the center to center distance of the two end semi-circles.
• Click to specify the width of the slot. Figure-26 shows the procedure of creating straight slot.

Figure-26. Straight Slot Creation

Centerpoint Straight Slot

- Click on the **Centerpoint Straight Slot** tool from the drop-down. The **Slot PropertyManager** will be displayed as discussed earlier. In this **PropertyManager**, the **Centerpoint Straight Slot** button is selected by default in the top section.
- Click to specify the center of the slot.
- Click to specify the half length of the center to center distance of the slot.
- Click to specify the length of the slot. Figure-27 shows the procedure of creating the centerpoint straight slot.

Figure-27. Centerpoint Straight Slot Creation

3 Point Arc Slot

- Click on the **3 Point Arc Slot** tool from the drop-down. The **Slot PropertyManager** will be displayed as earlier. In this **PropertyManager** the **3 Point Arc Slot** button is selected by default in the top section.
- Click to specify the start point of the center arc.
- Click to specify the end point of the center arc.
- Click to specify the radius of the slot.
- Click to specify the width of the slot.

Centerpoint Arc Slot

- Click on the **Centerpoint Arc Slot** tool from the drop-down. The **Slot PropertyManager** will be displayed as earlier. In this **PropertyManager** the **Centerpoint Arc Slot** button is selected by default in the top section.
- Click to specify the center of the slot center arc.
- Click to specify the start point of the center arc.
- Click to specify the end point of the slot.
- Click to specify the width of the slot.

Circle

There are two tools in this drop-down; **Circle** and **Perimeter Circle**; refer to Figure-28.

Figure-28. Circle drop-down

The tools in this drop-down are explained next.

Circle

- Click on the **Circle** tool from the drop-down. The **Circle PropertyManager** will be displayed; refer to Figure-29.

Figure-29. Circle Property-Manager

- Click to specify the center point of the circle.
- Click to specify the diameter of the circle. If you want to specify the radius of the circle then clear the **Diameter dimensions** check box from the **PropertyManager** after selecting the **Circle** tool.

Perimeter Circle

- Click on the **Perimeter Circle** tool from the drop-down. The **Circle PropertyManager** will be displayed as earlier. In this **PropertyManager**, the **Perimeter Circle** button is selected by default.
- Click one by one at three locations to specify three perimeter points through which the circle should pass.

Arc

There are three tools in this drop-down; **Centerpoint Arc**, **Tangent Arc**, **3 Point Arc**; refer to Figure-30.

Figure-30. Arc drop-down

The tools in this drop-down are explained next.

Centerpoint Arc

- Click on the **Centerpoint Arc** tool from the drop-down. The **Arc PropertyManager** will be displayed as shown in Figure-31.
- Click to specify the center point of the arc.
- Click to specify the start point of the arc.
- Click to specify the end point of the arc. Refer to Figure-32 for the creation of the arc.

Figure-31. Arc PropertyManager

Figure-32. Arc Creation

Tangent Arc

- Click on the **Tangent Arc** tool from the drop-down. The **Arc PropertyManager** will be displayed as earlier. In this **PropertyManager**, the **Tangent Arc** button is selected by default.
- Select end point of the entity to which the arc should be tangent.
- Click to specify the end point of the arc. Note that the tangent constraint is automatically applied at the arc.
- From SolidWorks 2018 onwards, you can reverse the direction of tangency at end points of arcs. To do so, right-click on arc. A shortcut menu will be displayed; refer to Figure-33. Select the **Reverse Endpoint Tangent** option from the menu to reverse the direction.

3 Point Arc

- Click on the **3 Point Arc** tool from the drop-down. The **Arc PropertyManager** will be displayed as earlier. In this **PropertyManager**, the **3 Point Arc** button is selected by default.
- Click to specify the start point of the arc.
- Click to specify the end point of the arc.
- Click to specify a point on the arc to set the radius of the arc.

Note that now you can detach a connected geometric entity by dragging. To do so, you need to activate the detach mode. Right-click on the entity that you want to be detached by dragging and select the **Detach Segment on Drag** option from the shortcut menu; refer to Figure-34. Now if you select and drag the geometric entity then it will get detached from the original section in sketch.

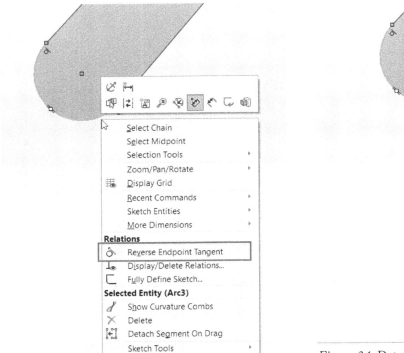

Figure-33. Shortcut menu for tangent arcs

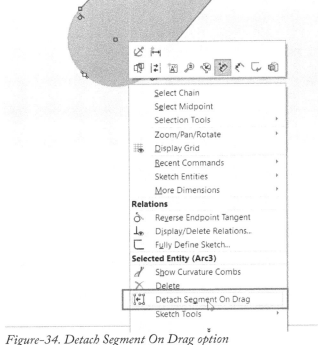

Figure-34. Detach Segment On Drag option

Polygon

The **Polygon** tool is used to create polygons of desired number of sides. The procedure to create polygon is explained next.

- Click on the **Polygon** tool from the **Ribbon**. The **Polygon PropertyManager** will be displayed as shown in Figure-35.
- Specify the number of sides of the polygon by using the spinner in the **Parameters** rollout.
- Select the desired radio button from the rollout. If you want to create the polygon inside the circle, then select the **Inscribed circle** radio button. If you select the **Circumscribed circle** radio button, the polygon will be drawn outside the circle. Note that polygon corner points lie on the circumscribed circle if you select the **Circumscribed circle** radio button. On the other hand, sides of polygon are tangent to the circle if you have selected the **Inscribed circle** radio button.
- Click to specify the center point of the circle.
- Click to specify the corner point of the polygon. Figure-36 shows the procedure of creating polygons.

Figure-35. Polygon PropertyManager

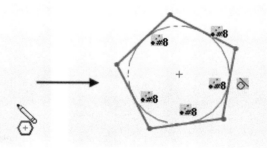

Figure-36. Polygon creation

Spline

The tools in **Spline** drop-down are used to create splines with different methods. Figure-37 show the tools of this drop-down. These tools are discussed next.

Figure-37. Spline drop-down

Spline

- Click on the **Spline** tool from the drop-down. You will be asked to specify the points through which the spline should pass.
- One by one click in the viewport to specify the points of the spline.
- To end specifying points, press the **ESC** key from the keyboard.

- Now, select the spline created to modify its parameters. The **Spline PropertyManager** will be displayed as shown in Figure-38.
- Now, you can use the options in the **PropertyManager** to modify the spline like, select the **Show curvature** check box in **Options** rollout to display curvature of spline.

Figure-38. Spline PropertyManager

- After performing the modifications, click on the **OK** button from the **PropertyManager**.
- If you want to add more control points on the spline, then right-click on the spline. A shortcut menu will be displayed; refer to Figure-39.
- Select the **Insert Spline Point** option from the shortcut menu. You will be asked to specify the location of spline point to be inserted.
- Click at the desired location(s) on the spline to create the point(s). Press ESC to exit the option.
- Similarly, you can add curvature control point or specify the other options in the shortcut menu.

Figure-39. Shortcut menu for spline

Style Spline

- Click on the **Style Spline** tool from the **Spline** drop-down. The **Insert Style Spline PropertyManager** will be displayed; refer to Figure-40. Also, you will be asked to specify the control points for the spline. (Control points control the shape of the spline but they do not lie on the spline).
- Select the desired radio button from the **Spline Type** rollout to set the spline style or say set the smoothness of spline.
- One by one click to specify the control points of the spline.
- To end specifying control points, press the ESC key from the keyboard.

*Figure-40. Insert Style Spline Property-
Manager*

Note that the minimum points required to create 3°, 5°, and 7° B-Splines are 4, 6, and 8 respectively. If you specify points lesser than required then you will be creating a curve rather than spline!!

Figure-41 shows the splines created.

Spline create using points

Spline created using the Control Points

Figure-41. Spline creation

Equation Driven Curve

The **Equation Driven Curve** tool is very helpful in creating curves using mathematical equations. Follow the procedure given next to create the curve using the equation.

- Click on the **Equation Driven Curve** tool from the **Spline** drop-down. The **Equation Driven Curve PropertyManager** will display as shown in Figure-42.

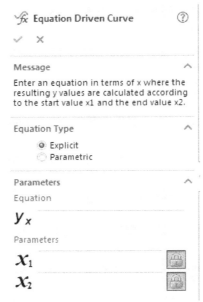

Figure-42. Equation Driven Curve PropertyManager

- Click on the desired radio button to create the **Explicit** or **Parametric** equation. **Explicit** means **Y** is function of **x** and **Parametric** means **X** and **Y** are functions of **t**.
- Specify the parameters for y_x. For example, in the Y_x edit box specify $x^2/10$.
- Click in the x1 edit box and specify the starting value. For example, specify the value as **0**.
- Click in the x2 edit box and specify the ending value. For example, specify the value as **20**.
- Figure-43 shows the output spline created by specifying the above parameters.

- If you have selected the **Parametric** radio button from the **PropertyManager** then specify the values of formulae for X_t and Y_t. Next, specify the limits in t_1 and t_2 edit boxes as discussed earlier.

Figure-43. Equation type

Ellipse

The tools in the **Ellipse** drop-down are used to create geometric profiles like; ellipse, partial ellipse, parabola, and conic; refer to Figure-44. The tools in the **Ellipse** drop-down are discussed next.

Figure-44. Ellipse drop-down

Ellipse

The **Ellipse** tool is used to create ellipses in the sketch. The procedure to create ellipse is discussed next.

- Click on the **Ellipse** tool. You are asked to specify the center of the ellipse.
- Click to specify the center. You are asked to specify the radius along the major axis.
- Click to specify the radius or specify it by entering value in the **PropertyManager**.
- Click to specify the radius along the minor axis. Figure-45 shows the process of creating ellipse. Once you click to specify the radius of minor axis, the **Ellipse PropertyManager** will be displayed; refer to Figure-46. Note that you can also specify the values in the edit boxes in **PropertyManager**.

Figure-45. Ellipse creation

Figure-46. Ellipse PropertyManager

Partial Ellipse

The **Partial Ellipse** tool is used to create partial ellipses in the sketch. The procedure to create partial ellipse is given next.

- Click on the **Partial Ellipse** tool from **Ellipse** drop-down. You will be asked to specify the center point of the ellipse.
- Click to specify the center point. You will be asked to specify the radius along major axis.
- Click to specify the radius. You will be asked to specify the radius along minor axis.
- Click to specify the radius along minor axis. The point where you specify the radius of minor axis will become the starting point of the partial ellipse.
- Click to specify the end point of the partial ellipse. Figure-47 shows the process of creating partial ellipse.
- After specifying the end point of the partial ellipse, the **Ellipse PropertyManager** displays as shown in Figure-48.

Figure-47. Partial ellipse creation

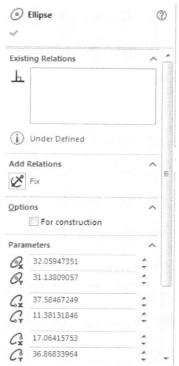

Figure-48. Partial Ellipse PropertyManager

Parabola

The **Parabola** tool is used to create parabola in the sketch. The procedure to create parabola is explained next.

- Click on the **Parabola** tool from the drop-down. You will be prompted to specify the focal center of the ellipse.
- Click to specify the focal center. You will be asked to specify the distance of focal center from the directrix.
- Click to specify the distance. You will be asked to specify the starting point of the parabola segment.
- Click on the dashed curve to specify the start point and then click to specify the end point of the parabola. Figure-49 shows the process of creating parabola.
- Once the parabola is created, the **Parabola PropertyManager** will be displayed; refer to Figure-50. You can change the parameters as per your requirement by using the options of **PropertyManager**.

Figure-49. Parabola creation

Figure-50. Parabola PropertyManager

Conic

The **Conic** tool is used to create conic curves in the sketch. The procedure to create a conic curve is explained next.

- Click on the **Conic** tool from the drop-down. You will be prompted to specify the start point of the conic.
- Click to specify the start point of the conic curve. You are asked to specify the end point of the conic curve.
- Click to specify the end point. You are asked to specify the top vertex of the curve.
- Click to specify the vertex. You are asked to specify the **Rho** value for the curve. Specify the desired value of **Rho**. Figure-51 shows the process of conic curve creation. Once the curve is created, the **Conic PropertyManager** is displayed as shown in Figure-52.

Figure-51. Conic creation

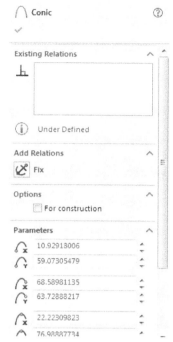

Figure-52. Conic PropertyManager

Sketch Fillet

The **Sketch Fillet** tool is used to create fillet at the corners created by intersection of two entities. Fillet is sometimes also referred to as round. Generally it is not advised to use this tool if there are some major changes going to occur later in the model. Once you have performed all the modifications then you apply fillets. The procedure to create fillet is given next.

- Click on the **Sketch Fillet** from the **Sketch Fillet** drop-down. The **Sketch Fillet PropertyManager** will be displayed; refer to Figure-53.

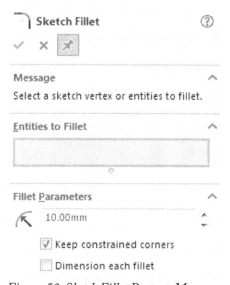

Figure-53. Sketch Fillet PropertyManager

- Enter the radius value in the **Fillet Radius** spinner of **Fillet Parameters** rollout. Select the first entity. You will be asked to select the second entity.

- Select the second entity. The preview of fillet is displayed.
- Click on the **OK** button from the **PropertyManager** or select the next two entities between which you want to create the fillet. Figure-54 show the process of fillet creation.

Figure-54. Fillet Creation

Sketch Chamfer

The **Sketch Chamfer** tool is used to create chamfer at the corners created by intersection of two entities. The procedure to create chamfers is explained next.

- Click on the **Sketch Chamfer** tool from the **Sketch Fillet** drop-down. The **Sketch Chamfer PropertyManager** will be displayed; refer to Figure-55.

*Figure-55. Sketch Chamfer
PropertyManager*

By default, **Distance-distance** radio button is selected in the **PropertyManager**. Specify the chamfer length for first side. If the **Equal distance** check box is selected then it will be applied for the both sides. If you clear the check box then you can specify the chamfer length for both sides by using the respective edit boxes.

You can also create chamfer by specifying angle and distance. To do so, select the **Angle-distance** radio button. Specify the parameters in the **Chamfer Parameters** rollout.

- Select the first line and then select the second line. The chamfer will be created between both the lines.

Text

The **Text** tool is used to create text which can be used for embossing/engraving on a solid face. The **Text** tool is also used to provide notes and other information for the model. The procedure to create text is explained next.

- Click on the **Text** tool. The **Sketch Text PropertyManager** will be displayed as shown in Figure-56.

Figure-56. Sketch Text PropertyManager

- Click in the **Text** box and enter the desired text you want to use in the sketch.
- The text will be created at the default coordinate system.
- Select a curve along which you want to create the text.
- The text will be placed along the curve. Figure-57 shows the procedure of creating text.

Figure-57. Text along curve

Note that if the curve length is smaller than the text specified then the text that can be spaced over the curve will only be displayed.

Point

The **Point** tool is used to create sketch point in the viewport. The point is a very important entity and finds its major use when you start creating surfaces. The points give the flexibility to parametrically change the surface design. The procedure to create points is explained next.

- Click on the **Point** tool from the **Ribbon**. You will be asked to click in the viewport to specify the location of the point.
- Click in the viewport or at desired location on a entity in viewport. The **Point PropertyManager** will be displayed as shown in Figure-58.

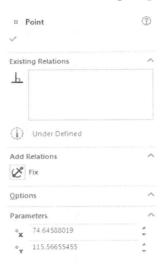

Figure-58. Point PropertyManager

- Click in the spinners of the **PropertyManager** to specify the parameters of the points.

SKETCH EDITING TOOLS

The standard tools to edit sketch entities are categorized in this section. The tools in this section are discussed next.

Trim Entities

The **Trim Entities** tool is available in the **Trim Entities** drop-down of the **Ribbon**. This tool is used to remove unwanted part of a sketch entity. While removing the segments, the tool considers intersection point as the reference for trimming. The procedure to use this tool is explained next.

- Click on the **Trim Entities** tool. The **Trim Entities PropertyManager** displays as shown in Figure-59.

By default, the **Power trim** button is selected. This button is having the functionality of all the other buttons displayed in the **PropertyManager**. So, we will be explaining the procedure of using this button first.

Figure-59. Trim PropertyManager

Power Trim

- Make sure the **Power Trim** button is selected in the **PropertyManager**.
- Click on the entity you want to trim. Note that the selected side of the entity will be removed.
- Click on the entity that you want to use as reference for trimming.
- The entity will be trimmed. Figure-60 shows the procedure of trimming.

Figure-60. Trimming procedure

Or

- Select the **Trim Entity** tool, make sure the **Power Trim** button is selected.
- Click and hold the mouse button, and drag the cursor over the portion of entities you want to be removed. Refer to Figure-61.

Figure-61. Trimming by dragging

Corner

The **Corner** tool is used to join two intersecting entities while forming corner at the intersection point. The procedure is given next.

* Click on the **Corner** button from the **Options** rollout of **Trim PropertyManager**. You will be asked to select the entities.
* Select the two intersecting sketch entities. The corner trim will be created; refer to Figure-62.

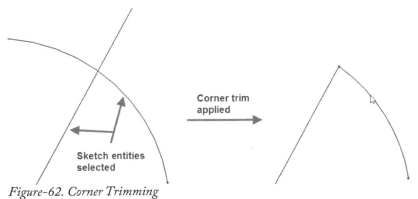

Figure-62. Corner Trimming

Trim away inside

The **Trim away inside** button is used to trim all the entities falling inside the selected boundaries. The procedure to use this option is given next.

* Click on the **Trim away inside** button from the **Trim PropertyManager**. You will be asked to select two bounding entities.
* Select a closed loop or select two bounding entities which are open loop like lines, arcs etc. You will be asked to select or box-selected the entities that are to be trimmed. The entities that are inside the selected boundaries will be trimmed; refer to Figure-63.

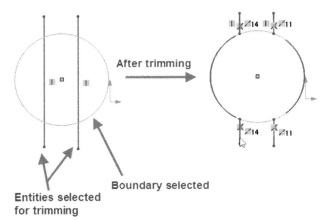

Figure-63. Using Trim away inside button

In the same way, you can use the **Trim away outside** and **Trim to closet** button in the **Trim PropertyManager**.

Extend Entities

The **Extend Entities** tool does the reverse of **Trim Entities** tool. This tool is available in the **Trim Entities** drop-down. This tool extends the sketch entities up to the nearest intersecting entity. The procedure to use this tool is discussed next.

- Click on the **Extend Entities** tool from the **Trim Entities** drop-down.
- Hover the cursor on the entity that you want to extend. Preview of the extension will display.
- Click on the entity if the preview is as per your requirement. If you want the entity to be extended in reverse direction then click on the other portion of the entity. Figure-64 shows the process of extending entities.

Figure-64. Extending process

The **Convert Entities** and **Intersection Curves** tools will be explained in the Modeling section of the book, later.

Offset Entities

The **Offset Entities** tool is used to create copy of the selected entities at a specified distance from the them. If you are the user of AutoCAD then this tool is the most common tool being used while creating layouts. The procedure to use this tool is given next.

- Click on the **Offset Entities** tool from the **Ribbon**. The **Offset Entities PropertyManager** will be displayed; refer to Figure-65.

Figure-65. Offset Entities PropertyManager

There are various options in this **PropertyManager** that are linked to the output of the tool. So, we will discuss them one by one.

Offset Distance

This spinner is used to set the distance value for offset. You can also enter the desired value in the edit box.

Add dimensions

Select this check box if you want to create the dimensions while creating offset.

Reverse

Select this check box if you want to reverse the direction of offset being displayed.

Select chain

Select this check box if you want to select the chain of entities connecting to the selected entity.

Bi-directional

Select this check box if you want to create the offset entities on both sides of the selected entity.

Construction Geometry

There are two check boxes in this section, **Base geometry** and **Offset geometry**, to make the base and offset geometry as construction entity. Select the desired check box/boxes from the section.

Cap ends

Select this check box if you want to close the ends of offset entities. This check box is active when the **Bi-directional** check box is selected and entity is selected. After selecting the **Cap ends** check box, you can close the offset entities by using the arcs or lines. For using arcs or lines, select the respective radio button.

- After selecting the desired options, select the entity from the viewport. Preview will be displayed in yellow color.
- Click on **OK** button to create the offset. Figure-66 shows the process of creating offset entities.

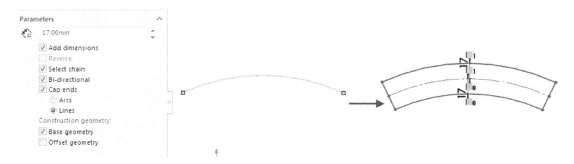

Figure-66. Offset entities creation

Mirror Entities

The **Mirror Entities** tool is used to create mirror copy of the selected entities with respect to a reference called mirror line. The procedure to create mirror entities is given next.

*   Click on the **Mirror Entities** tool from the **Ribbon**. The **Mirror PropertyManager** is displayed; refer to Figure-67.

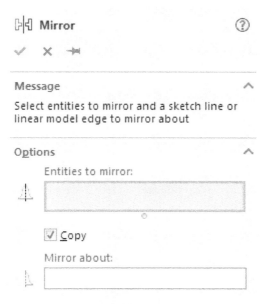

Figure-67. Mirror PropertyManager

*   Select the entity/entities you want to create mirror copy of.
*   Deselect the **Copy** check box from the **PropertyManager** if you want to delete the original entities after creating mirror copy.
*   Click in the **Mirror about** box and select the reference line that you want to use as mirror line. The mirror line can be an edge of a solid, sketch line, or a centerline. Figure-68 shows the process of creating mirror entities.

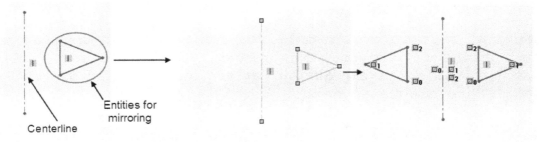

Figure-68. Mirror entities creation

Dynamic Mirror

The Dynamic Mirror tool is used to create a mirror wall so that geometries created on one side of mirror line are also created on other side as mirrored object. The procedure to enable dynamic mirror is given next.

- Click on the **Dynamic Mirror** tool from the **Sketch Tools** cascading menu of the **Tools** menu. You will be asked to select a sketched line or linear edge about which the entities will be mirrored. Also, the Mirror PropertyManager will be displayed; refer to Figure-69.

Figure-69. Mirror PropertyManager

- Select a line or edge. The line will be converted as mirror line.
- Now, if you create any entity on one side of the mirror line then a reflection will be created on the other side; refer to Figure-70.

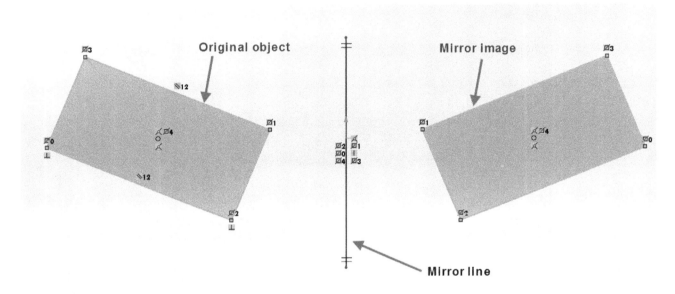

Figure-70. Dynamic mirror sketching

- To exit dynamic mirror mode, click again on **Dynamic Mirror** tool from the **Sketch Tools** cascading menu of the **Tools** menu.

Sometimes, we need to create multiple copies of the sketch entities. Like in sketch of a keyboard or piano. For such cases, SolidWorks provides two tools, **Linear Sketch Pattern** and **Circular Sketch Pattern**. These tools are discussed next.

Linear Sketch Pattern

The **Linear Sketch Pattern** tool is used to create multiple copies of an entity in linear directions. You can create pattern in two linear directions at a time. The procedure to create linear sketch pattern is given next.

- Click on the **Linear Sketch Pattern** tool from the **Linear Sketch Pattern** drop-down of the **Ribbon**. The **Linear Pattern PropertyManager** will be displayed as shown in Figure-71.

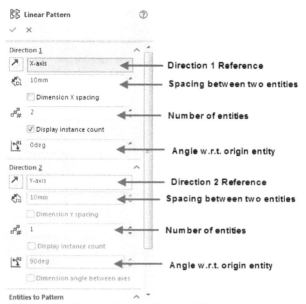

Figure-71. Linear Pattern PropertyManager

- Select the entity that you want to pattern. Specify the parameters as per your requirement.
- If you want to create pattern along an axis or line then click in the **Direction 1** reference box or **Direction 2** reference box as per your need. Now, select the desired axis or line to specify the direction reference.
- If you want to skip any of the entity created in pattern then expand the **Instances to skip** rollout at the bottom of the **PropertyManager** and click in the **Instances to skip** box. Now, click on the pink dot for the entities from the preview that you do not want to create; refer to Figure-72.

Figure-72. Instances to be removed

- Increase the number of entities in the **Direction 2** rollout to activate the options in the rollout. Figure-73 shows the pattern created for a circle and its respective options in the **PropertyManager**.

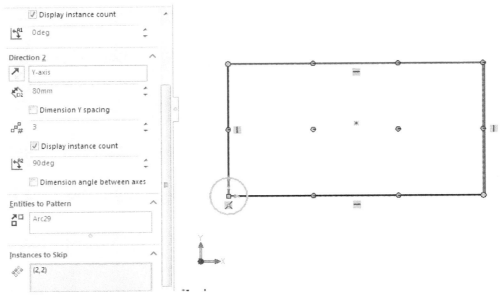

Figure-73. Linear pattern creation

Circular Sketch Pattern

The **Circular Sketch Pattern** tool is used to create multiple copies of an entity in circular fashion. The procedure to create circular sketch pattern is given next.

- Click on the **Circular Sketch Pattern** tool from the **Linear Sketch Pattern** drop-down. The **Circular Pattern PropertyManager** will be displayed; refer to Figure-74.

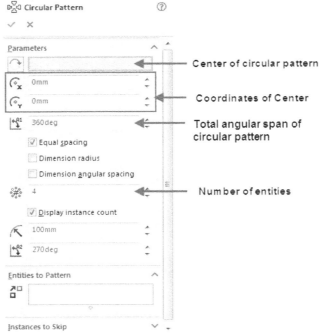

Figure-74. Circular Pattern PropertyManager

- Select the entity that you want to pattern and click in the **Center of circular** selection box at the top in the **PropertyManager** to specify the center of the pattern.
- Select the point to be the origin, specify the number of entities and click on the **OK** button to create the pattern.

- You can skip the entities as you did for Linear pattern. Figure-75 shows a circular pattern created with its parameters specified.

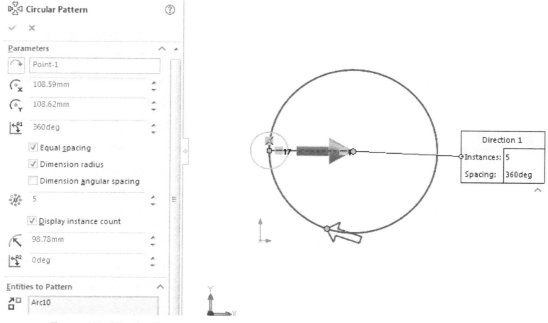

Figure-75. Circular Pattern creation

Move Entities

The **Move Entities** tool is used to move the entities from one position to another position. You can move the entities either by coordinate values or by clicking. The procedure to move entities is discussed next.

- Select the **Move Entities** tool from the **Ribbon**. The **Move Entities PropertyManager** will be displayed as shown in Figure-76.
- Select the entity that you want to move.
- Make sure **From/To** radio button is selected. Now, click on the point you want to use as base point.
- Click at the destination point where you want to place the entity. The entity will move to specified place.

Figure-76. Move Entities PropertyManager

Or

- Click on the **X/Y** radio button and enter the distance in X and Y direction to place the sketch entity.

Or

- Select the entity, you want to move and drag it from the key point/ curve. Note that circular entities like circle, arc, ellipse are dragged from center for moving them. For the other entities, select the curve from the location which is not a key point. Figure-77 shows a spline and line moved together by dragging.

Figure-77. Moving entities by dragging

Copy Entities

The **Copy Entities** tool is available in the **Move Entities** drop-down. This tool is used to copy the entities by specifying position. This tool works in the same way as the **Move Entities** tool does. The only difference is that it does not move the entity but it creates the entities. Figure-78 shows the entities copied by this tool.

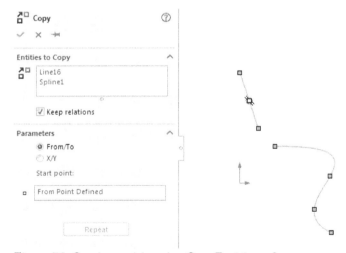

Figure-78. Copying entities using Copy Entities tool

Rotate Entities

The **Rotate Entities** tool is available in the **Move Entities** drop-down. This tool is used to rotate the entities by specifying angle. The procedure to use this tool is given next.

- Click on the **Rotate Entities** tool from the drop-down. The **Rotate PropertyManager** will be displayed as shown in Figure-79.

Figure-79. Rotate PropertyManager

- Select the entity that you want to rotate.
- Click in the **Center of rotation** box and click on the point that you want to make center of rotation.
- Specify angle in the **Angle** spinner. The object will rotate by specified value. Refer to Figure-80.

Figure-80. Rotating an entity

Scale Entities

The **Scale Entities** tool is available in the **Move Entities** drop-down. This tool is used to increase or decrease the size of an entity by specified scale value. The procedure to use this tool is given next.

- Click on the tool from the drop-down. The **Scale PropertyManager** will display as shown in Figure-81.

Figure-81. Scale PropertyManager

- Select the entities that you want to scale up or scale down.
- Click in the **Scale about** box and select the base point about which you want to scale the entities.
- Enter the scale value in the **Scale Factor** spinner. You can create a copy of entities scaled to specified value by selecting the **Copy** check box. In that case, the original entities will remain unchanged; refer to Figure-82.

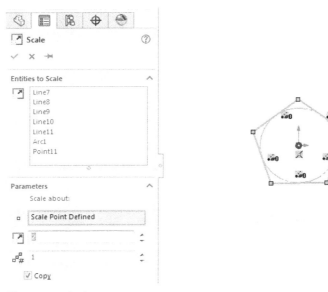

Figure-82. Scaling entities

Stretch Entities

The **Stretch Entities** tool is available in the **Move Entities** drop-down. This tool is used to stretch any sketched entity. The procedure to use this tool is given next.

- Click on the **Stretch Entities** tool from the drop-down. The **Stretch PropertyManager** will be displayed as shown in Figure-83.

Figure-83. Stretch PropertyManager

- By using the cross-rectangle selection, select the portion of entities that you want to stretch.
- Click in the **Stretch about** selection box and then select the base point on the sketch.
- Move the cursor to the desired position and click to stretch the entities. Figure-84 shows the procedure of stretching the entities.

Figure-84. Stretching entities

Segment Tool

The **Segment** tool is used to create segments of circles, arcs, spline, parabolas etc. By default, the **Segment** tool is not available in the **CommandManager** and you need to add the tool manually. The procedure to add the tool to **CommandManager** is given next.

Adding Segment Tool in CommandManager

- Right-click on any tool in the **CommandManager** and select the **Customize** option from the shortcut menu displayed; refer to Figure-85. The **Customize** dialog box will be displayed; refer to Figure-86.

Figure-85. Customize option in shortcut menu

Figure-86. Customize dialog box

- Click on the **Commands** tab in the dialog box and select the **Sketch** option from the **Categories** area. The buttons of the **Sketch** categories will be displayed; refer to Figure-87.

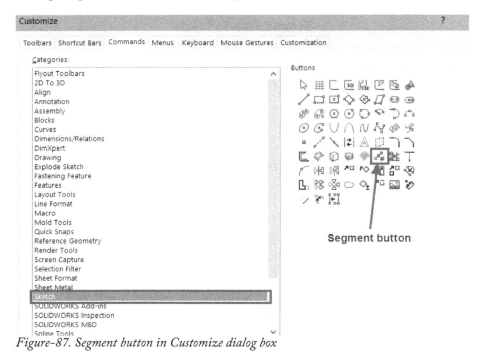

Figure-87. Segment button in Customize dialog box

- Select the **Segment** button from the dialog box and drag-drop it in the **Sketch CommandManager**; refer to Figure-88.

Figure-88. Dropping Segment button in CommandManager

- Click on the **OK** button from the dialog box.

Creating Segments of Sketch Entity

- Click on the newly added **Segment** tool from the **Sketch CommandManager**. The **Segment PropertyManager** will be displayed; refer to Figure-89.

Figure-89. Segment Property-Manager

- Select the entity that you want to break into equal segments. Preview of segment will be displayed; refer to Figure-90.

Figure-90. Preview of segment

- By default, the **Sketch points** radio button is selected in the **PropertyManager** and hence the points are created on the line at equal distance. Select the **Sketch segments** to create segments of entity in place of only points.
- Set the number of segments in the spinner displayed in **PropertyManager**.
- Click on the **OK** button from the **PropertyManager** to create the segments/points.

Split Entities

The **Split Entities** tool is used to split the selected sketch entities at specified points. By default this tool is not available in the **CommandManager** so you need the add this tool in the same way as you have added the **Segment** tool in the **Sketch CommandManager**. The procedure to use this tool is given next.

- Click on the newly added **Split Entities** tool from the **Sketch CommandManager**. The **Split Entities PropertyManager** will be displayed; refer to Figure-91. Also, you will be asked to select a sketch entity to be split.

Figure-91. Split Entities PropertyManager

- Click at the desired location on the sketch entity to split it at that location.
- Specify as many split points as you want and then press ESC from keyboard.

RELATIONS

Relations are used to constrain the sketch entities dimensionally and/or geometrically. In SolidWorks also you can constrain any sketch entity by using dimensions or geometrical constrains. Both type of constrains are explained next.

Dimensional Constraints (Dimensions)

Dimensions are used to limit size of the sketch entities. For example; specifying length of the line, specifying diameter of the circle, and so on. The tools to dimension sketch entities are given in the **Smart Dimension** drop-down. The tools that are commonly used for dimensioning are discussed next.

Smart Dimension

The **Smart Dimension** tool is used to dimension the entities automatically. This tool can create different type of dimensions like, horizontal, vertical, or inclined dimensions. The procedure to create dimensions are explained next.

- Click on the **Smart Dimension** tool from the **Smart Dimension** drop-down in the **Ribbon**. You are asked to select entities to dimension.
- Select the entity you want to dimension and then click at the desired distance to place dimension.

Selection pattern for dimensioning various entities after selecting the **Smart Dimension** tool is given next.

Dimensioning a line

Click on the line to be dimensioned and then click at desired position to place the dimension.

Dimensioning inclined line

Click one by one at the end points of the line. Move the cursor perpendicularly above the line to create inclined dimensions. If you want create horizontal dimension of a line then move the cursor vertically downward or upward beyond the limit of the line. Similarly, move the cursor towards left or right of the line to create vertical dimension.

Dimensioning arcs/circle

Click on the arc/circle to specify its radius/diameter.

Dimensioning elliptical arcs/ellipse

Click on the end points of the elliptical arcs/ellipses to dimension it.

Figure-92 show various entities dimensioned.

Figure-92. Dimensioning

Ordinate Dimension

You can also use Ordinate dimensioning if you are dimensioning for CNC coordinates. The procedure to create ordinate dimensions is given next.

- Click on the **Ordinate Dimension** tool from the **Smart Dimension** drop-down.
- Select the first reference that you want to make zero reference.
- Click to place the zero reference.
- Select the next line for which you want to specify the dimension.
- After you have specified the dimensions in one direction. Press the **ESC** key and then restart the tool.

- Now, select the next zero reference and repeat the procedure. Figure-93 shows a sketch dimensioned by ordinates.

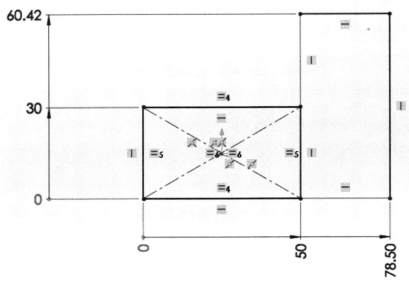

Figure-93. Ordinate dimensioned sketch

Note that you can use the **Horizontal Ordinate Dimension** and **Vertical Ordinate Dimension** tools in the **Dimension** drop-down in the same way.

Path Length Dimension

The **Path Length Dimension** tool is used to dimension the total length of selected entities. The procedure to use this tool is given next.

- Click on the **Path Length Dimension** tool from the **Smart Dimension** drop-down in the **Sketch CommandManager** of **Ribbon**. The **Path Length PropertyManager** will be displayed and you will be asked to select the sketch entities.
- Select the curve(s) whose path length is to be dimensioned and then click on the **OK** button from the **PropertyManager**. The Path Length dimension will be attached to the curve(s); refer to Figure-94.

Note that the information about path length, angle, distance etc. is also displayed in the status bar on selecting the entities; refer to Figure-95.

Figure-94. Path length dimension created

Figure-95. Information displayed in status bar

Geometric Constraints

These constraints are used to constrain the shape/position of sketch entities with respect to other entities.

• To apply the geometric constraints, click on the **Add Relation** button in the **Display/Delete Relations** drop-down.

The list of constraints that can be applied in SolidWorks is given next.

Horizontal

The **Horizontal** constraint makes one or more selected lines or center lines to become horizontal. You can also select an external entity such as an edge, plane, axis, or sketch curve on an external sketch that will act as a line to apply this constraint. You can also make two or more points to become horizontal using the **Horizontal** constraint. A point can be a sketch point, a center point, an endpoint, a control point of a spline, or an external entity such as origin, vertex, axis, or point in an external sketch. To apply this constraint, invoke the **Add Relations PropertyManager**. Select the entity or entities to apply the **Horizontal** constraint. Choose the **Horizontal** button from the **Add Relations** rollout in the **Add Relations PropertyManager**. You will notice that the name of the horizontal constraint will be displayed in the **Existing Constraints** rollout.

Vertical

The **Vertical** constraint makes one or more selected lines or centerlines to become vertical. You can force two or more points to become vertical using the **Vertical** constraint. To apply this constraint, invoke the **Add Relations PropertyManager** and select the entity or entities to apply the **Vertical** constraint. Choose the **Vertical** button from the **Add Relations** rollout. You will notice that the name of the vertical constraint is displayed in the **Existing Constraints** rollout.

Collinear

The **Collinear** constraint makes the selected lines to lie on the same infinite line; refer to Figure-96. To use this constraint, select the lines to apply the **Collinear** constraint. Choose the **Collinear** button from the **Add Relations** rollout.

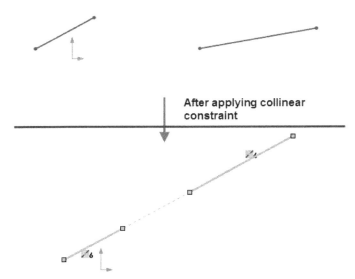

Figure-96. Applying collinear constraint

Coradial

The **Coradial** constraint makes the selected arcs or circles to share the same radius and the same center points; refer to Figure-97. You can also select an external entity that projects as an arc or a circle in the sketch to apply this constraint. To use this constraint, invoke the **Add Relations PropertyManager**. Select two arcs or circles, or an arc and a circle to apply the **Coradial** constraint. Choose the **Coradial** button from the **Add Relations** rollout.

Figure-97. Applying coradial constraint

Perpendicular

The **Perpendicular** constraint makes the selected lines to become perpendicular to each other. To apply this constraint, invoke the **Add Relations PropertyManager**. Select two lines and choose the **Perpendicular** button from the **Add Relations** rollout. Figure-98 shows two lines before and after applying the **Perpendicular** constraint.

Figure-98. Applying perpendicular constraint

Parallel

The **Parallel** constraint makes the selected lines to become parallel to each other. To apply this constraint, invoke the **Add Relations PropertyManager**. Select two lines and choose the **Parallel** button from the **Add Relations** rollout.

ParallelYZ

The **ParallelYZ** constraint makes a line in the three-dimensional (3D) sketch to become parallel to the YZ plane with respect to the selected plane. To apply this constraint, invoke the **Add Relations PropertyManager**. Select a line in the 3D sketch and then select a plane. Next, choose the **ParallelYZ** button from the **Add Relations** rollout.

ParallelZX

The **ParallelZX** constraint makes a line in the 3D sketch to become parallel to the ZX plane with respect to the selected plane. To apply this constraint, invoke the **Add Relations PropertyManager**. Select a line in the 3D sketch and then select a plane. Next, choose the **ParallelZX** button from the **Add Relations** rollout.

Along X

The **AlongX** constraint makes a line in the 3D sketch to become parallel to the X-axis. To apply this constraint, invoke the **Add Relations PropertyManager**. Select a line in the 3D sketch and then choose the **Along X** button from the **Add Relations** rollout; the selected line will be oriented along the X axis.

Along Y

The **Along Y** constraint makes a line in the 3D sketch to become parallel to the Y-axis. To apply this constraint, invoke the **Add Relations PropertyManager**. Select a line in the 3D sketch and then choose the **Along Y** button from the **Add Relations** rollout; the selected line will be oriented along the Y axis.

Along Z

The **AlongZ** constraint makes a line in the 3D sketch to become parallel to the Z-axis. To apply this constraint, invoke the **Add Relations PropertyManager**. Select a line in the 3D sketch and then choose the **Along Z** button from the **Add Relations** rollout; the selected line will be oriented along the Z axis.

Normal

The **Normal** constraint makes a line in the 3D sketch to become normal to the selected plane; refer to Figure-99. To apply this constraint, invoke the **Add Relations PropertyManager**. Select a line in the 3D sketch and then select a plane. Next, choose the **Normal** button from the **Add Relations** rollout; the selected line will be oriented normal to the plane.

Figure-99. Applying Normal constraint

On Plane

The **On Plane** constraint makes a line in the 3D sketch to become parallel and places on the selected plane. To apply this constraint, invoke the **Add Relations PropertyManager**. Select a line in the 3D sketch and then select a plane. Next, choose the **On Plane** button from the **Add Relations** rollout; the selected line will be oriented parallel to the selected plane and places on it.

Tangent

The **Tangent** constraint makes a selected arc, circle, spline, or ellipse to become tangent to other arc, circle, spline, ellipse, line, or edge. To apply this constraint, invoke the **Add Relations PropertyManager**. Select two entities and then choose the **Tangent** button from the **Add Relations** rollout.

Concentric

The **Concentric** constraint makes a selected arc or circle to share the same center point with other arc, circle, point, vertex, or circular edge. To apply this constraint, invoke the **Add Relations PropertyManager**. Select the required entity to apply the **Concentric** constraint and then choose the **Concentric** button from the **Add Relations** rollout.

Equal

The **Equal** constraint makes the selected lines to have equal length and the selected arcs, circles, or arc and circle to have equal radii. To apply this constraint, invoke the **Add Relations PropertyManager**. Select the required entity to apply the **Equal** constraint and choose the **Equal** button.

Intersection

The **Intersection** constraint makes a selected point to move at the intersection of two selected lines. To apply this constraint, invoke the **Add Relations PropertyManager**. Select the required entity to apply the **Intersection** constraint. Choose the **Intersection** button from the **Add Relations** rollout.

Coincident

The **Coincident** constraint makes a selected point to be coincident with a selected line, arc, circle, or ellipse. To apply this constraint, invoke the **Add Relations PropertyManager**. Select the required entity to apply the **Coincident** constraint. Choose the **Coincident** button from the **Add Relations** rollout.

Midpoint

The **Midpoint** constraint makes a selected point to move to the midpoint of a selected line. To apply this constraint, invoke the **Add Relations PropertyManager**. Select the point and the line to which the midpoint constraint has to be applied. Choose the **Midpoint** button from the **Add Relations** rollout.

Symmetric

The **Symmetric** constraint makes two selected lines, arcs, points, and ellipses to remain equidistant from a centerline. This constraint also makes the entities to have the same orientation. To apply this constraint, invoke the **Add Relations PropertyManager**. Select the required entity to apply the **Symmetric** constraint and select a center line. Choose the **Symmetric** button from the **Add Relations** rollout.

Fix

The **Fix** constraint makes the selected entity to be fixed at the specified position. If you apply this constraint to a line or an arc, its location will be fixed but you can change its size by dragging the endpoints. To apply this constraint, invoke the **Add Relations PropertyManager**. Select the required entity and choose the **Fix** button.

Merge

The **Merge** constraint makes two sketch points or endpoints to merge in a single point. To apply this constraint, invoke the **Add Relations PropertyManager**. Select the required entities to apply the **Merge** constraint and choose the **Merge** button from the **Add Relations** rollout.

Pierce

The **Pierce** constraint makes a sketch point or an endpoint of an entity to be coincident with an entity of another sketch. To apply this constraint, invoke the **Add Relations PropertyManager**. Select the required entities to apply the **Pierce** constraint and choose the **Pierce** button from the **Add Relations** rollout.

Fully Defined Sketch

Fully defined sketch is the one whose specifications can not be changed unintentionally. In some complex sketches, when you change dimension of one entity, the dimension of other entity gets changed automatically.

If you have the sketch fully defined then the dimension of the entities will not change unintentionally. In technical terms, a fully defined sketch is the one in which entities have zero degree of freedom. In SolidWorks, the sketch that is fully defined will be displayed in bold black color; refer to Figure-100.

Figure-100. A fully defined sketch

You can fully define a sketch either by manually applying the dimensions and geometric constraints or you can do it by using the **Fully Define Sketch** button. This button is available in the **Display/Delete Relations** drop-down. The procedure to use this tool is explained next.

- Click on the **Fully Defined Sketch** button in the **Display/Delete Relations** drop-down of the **Ribbon**. The **Fully Define Sketch PropertyManager** will display as shown in Figure-101. Note that all the rollouts in the **PropertyManager** are expanded.

- Select the type of dimensions and constraints that you want to use while making a sketch fully defined.
- Click on the **Calculate** button and then click on **OK** button to fully define the sketch. The dimensions and constraints will be automatically applied. Now, change the values as per your requirement. Refer to Figure-102.

Figure-102. Fully defined sketch

Figure-101. Fully Define Sketch PropertyManager

If you add dimensions or constraints that are more than required then such sketches are called over defined sketches. The dimensions, constraints and sketches that are over defined are displayed in yellow color. Refer to Figure-126. In such cases, you need to delete the conflicting dimensions or make them driving. You can also use the **SketchXpert** to do the modifications. The procedure is given as follow:

- Click on the **Over Defined** message in the Information Bar at the bottom of the viewport. Refer to Figure-103.

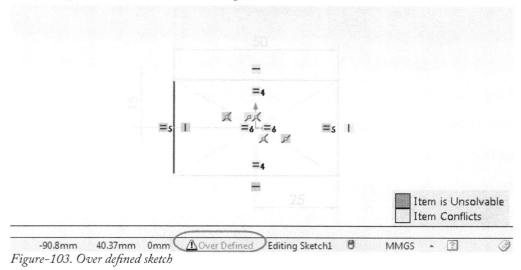

Figure-103. Over defined sketch

- On clicking the option, the **SketchXpert PropertyManager** will be displayed as shown in Figure-104.

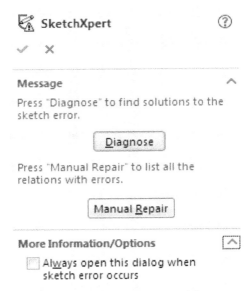

Figure-104. SketchXpert PropertyManager

- Click on the **Diagnose** button to find automatic solutions. The modified **SketchXpert PropertyManager** will display and interfering dimension will be displayed crossed; refer to Figure-105.

Figure-105. Diagnosed sketch

- Click on the arrow button to check the next solution and once you find the appropriate solution, click on the **Accept** button from the **Results** rollout.
- Click on **OK** button from the **PropertyManager** to accept the changes.

Till this point, we have completed the 2D sketching of SolidWorks. In next chapter, we will practice on the tools discussed till now using some examples.

SELF ASSESSMENT

Q1. Discuss about Planes and their placement in 3D space in the classroom. Use the example of a closed room to define different planes.

Q2. Which of the following is not the default place available in SolidWorks?

a. Front Plane
b. Right Plane
c. Top Plane
d. Left Plane

Q3. Which of the following is selected by default in the **Insert Line PropertyManager** while creating a line?

a. As sketched
b. Horizontal
c. Vertical
d. Angle

Q4. How many tools are available in SolidWorks to create a rectangle?

a. 3
b. 6
c. 5
d. 4

Q5. Which of the following tool is not available in SolidWorks to create an arc?

a. Tangent Arc
b. 3 Point Arc
c. Centerpoint Arc
d. 2 Point and Angle Arc

Q6. If you select the **Circumscribed circle** radio button while creating polygon then, the polygon will be drawn outside the circle. (T/F)

Q7. If you want to create the polygon inside the circle, then select the **Circumscribed circle** radio button from the **PropertyManager**. (T/F)

Q8. The **Equation Driven Curve** tool is available in the...........drop-down.

Q9. Discuss the use of **Trim away inside** button in the **Trim PropertyManager**.

Q10. Discuss the difference between Linear pattern and circular pattern.

Q11. Discuss the use of **Smart Dimension** tool with example.

Q12. What is the difference between dimensional constraining and geometrical constraining?

SPACE FOR STUDENT NOTES

Chapter 3

Advanced Dimensioning and Practice

Topics Covered

The major topics covered in this chapter are:

- *Dimensioning and its relations with drawing.*
- *Dimension Style.*
- *Practical 1.*
- *Practical 2.*
- *Practice Drawings*

DIMENSIONING AND ITS RELATIONS

When we dimension in a sketch in Sketching environment, it is not confined only to that environment but these dimensions and dimension styles are also fetched in Drafting and MBD environment of SolidWorks. You will learn later that the parameters specified in Sketching also affect the parameters in Drafting environment. At that time, the style and dimensions that we applied in sketch will be reflected in the draft. So, it is very important to understand dimension styles here. In the previous chapters, we have worked on basic dimensions. In this chapter, we will explain the dimensions and styles in detail.

DIMENSION STYLE

Select a dimension that you have applied in the sketch. The **Dimension PropertyManager** will be displayed; refer to Figure-1.

Figure-1. Dimension PropertyManager

The options in the **PropertyManager** are discussed next.

Style Rollout

The **Style** rollout is used to create, save, delete, and retrieve the dimension style in the current document; refer to Figure-2. You can also use the dimension styles saved in other documents using this rollout. The options in this rollout are discussed next.

Figure-2. The Style rollout

Apply the default attributes to selected dimensions

The **Apply the default attributes to selected dimensions** button is used to apply the default attributes to the selected dimension or dimensions. The attributes can be dimension text, tolerance, precision, arrow style and so on.

Add or Update a Style

The **Add or Update a Style** button is used to add a dimension style to the current document for a selected dimension. After invoking the **Dimension PropertyManager**, set the attributes using various options provided in this **PropertyManager**. Next, choose the **Add or Update a Style** button. The **Add or Update a Style** dialog box will be displayed, as shown in Figure-3. Enter the name of the dimension style in the edit box and press ENTER; the dimension style will be added to the current document.

You can apply a new dimension style to the selected dimension by selecting a dimension style from the **Set a current Style** drop-down list in the **Style** rollout. You can also update a dimension style. To do so, select the dimension and set the options of the dimension style according to your need. Next, choose the **Add or Update a Style** button to invoke the **Add or Update a Style** dialog box. Select the dimension style to update from the drop-down list in the dialog box; the two radio buttons in this dialog box will be enabled. Select the **Update all annotations linked to this Style** radio button and choose the **OK** button to update all the dimensions linked to the selected **Style**. If you select the **Break all links to this Style** radio button and choose the **OK** button, then the link between the other dimensions having the same style and the selected **Style** will be broken.

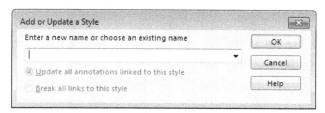

Figure-3. Add or Update a Style dialog box

Delete a Style

The **Delete a Style** button is used to delete a dimension style. Select a dimension style from the **Set a current Style** drop-down list and then choose the **Delete a Style** button.

Save a Style

The **Save a Style** button is used to save a dimension style so that it can be retrieved in some other document. Select a dimension style from the **Set a current Style** drop-down list and choose the **Save a Style** button. The **Save As** dialog box will be displayed. Browse to the folder in which you want to save the style and enter its name in the **File name** edit box. Choose the **Save** button from the **Save As** dialog box. The style file will be saved with the extension **.sldstl.**

Load Style

The **Load Style** button is used to open a saved style in the current document. The properties of that favorite will be applied to the selected dimension. To load a style, choose the **Load Style** button to invoke the **Open** dialog box. Browse to the folder in which the style is saved. Now, select the file with the extension **.sldstl** and choose the **Open** button; the **Add or Update a Style** dialog box will be displayed. Choose the **OK** button from this dialog box.

Tolerance/Precision Rollout

The **Tolerance/Precision** rollout shown in Figure-4 is used to specify tolerance and precision in dimensions. The options in this rollout are discussed next.

Figure-4. Tolerance/Precision rollout

Tolerance Type

The **Tolerance Type** drop-down is used to apply tolerance to a dimension. By default, the **None** option is selected. Therefore, no tolerance is applied to the dimensions. The other tolerance types available in this drop-down list are discussed next.

Basic
The basic dimension is the dimension taken as reference for other features. To display the basic dimension, select the dimension that you want to display as the basic dimension and then select the **Basic** option from the **Tolerance Type** drop-down list. On doing so, the dimension is enclosed in a rectangle; refer to Figure-5.

Bilateral
The **Bilateral** option is used to display the minimum and maximum limit of tolerance for a dimension. To apply the bilateral tolerance, select the dimension and then select the **Bilateral** option from the

Tolerance Type drop-down; the **Maximum Variation** and **Minimum Variation** edit boxes will be enabled, where you can enter the maximum and minimum variations for a dimension. Also, the **Show parentheses** check box will be displayed. If you select this check box, the bilateral tolerance will be displayed with parentheses. The dimension with a bilateral tolerance is shown in Figure-6.

Figure-5. Basic dimensions

Figure-6. Bilateral tolerance

Limit

The maximum and minimum permissible dimensional values of an entity are displayed on selecting the **Limit** option. To apply this tolerance type, select the dimension to be displayed as the limit dimension and select the **Limit** option; the **Maximum Variation** and **Minimum Variation** edit boxes will be enabled. Enter the values of the maximum and minimum variations. The dimension along with the limit tolerance is shown in Figure-7.

Symmetric

The symmetric tolerance when the variation is equal in both positive and negative direction. To use this tolerance, first select the dimension and then select the **Symmetric** option; the **Maximum Variation** edit box will be displayed. Enter the value of the tolerance in this edit box. Also, you can select the **Show parentheses** check box to show the tolerance in parentheses. The dimension along with the symmetric tolerance is shown in Figure-8.

Figure-7. Limit tolerance

Figure-8. Symmetric tolerance

MIN

In this dimensional tolerance, the **MIN.** symbol is added to dimension as suffix. This implies that the dimensional value is the minimum value that is allowed in the design. To display this dimensional tolerance, select a dimension and then the **MIN** option from the **Tolerance Type** drop-down list. The dimension along with the minimum tolerance is shown in Figure-9.

MAX

In this dimensional tolerance, the **MAX.** symbol is added to dimension as suffix. This implies that the dimensional value is the maximum value that is allowed in the design. To display this dimensional tolerance, select a dimension and then the **MAX** option from the **Tolerance Type** drop-down list. The dimension along with the maximum tolerance is shown in Figure-9.

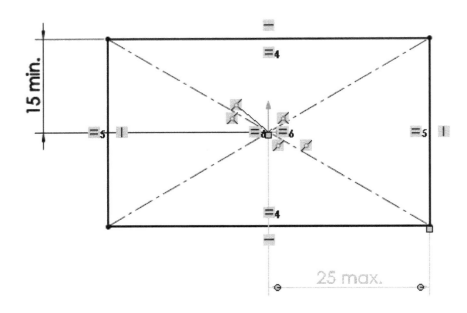

Figure-9. Min max tolerance

Fit

The **Fit** option is used to apply fit according to the Hole Fit and Shaft Fit systems. The **Tolerance/Precision** rollout with the **Fit** option selected in the **Tolerance Type** drop-down list is shown in Figure-10. Select the type of fit from the **Classification** drop-down list. The **Classification** drop-down list is used to define the **User Defined** fit, **Clearance** fit, **Transitional** fit, or **Press** fit. To apply a fit using the Hole Fit system or the Shaft Fit system, select the dimension and then the **Fit** option from the **Tolerance Type** drop-down list. The **Classification**, **Hole Fit**, and **Shaft Fit** drop-down lists will be displayed below the **Tolerance Type** drop-down list. Select the required fit from the **Classification** drop-down list and select the fit standard from the **Hole Fit** drop-down list or the **Shaft Fit** drop-down list. If you select the **Clearance**, **Transitional**, or **Press** option from the **Classification** drop-down list and the fit standard from the **Hole Fit** drop-down list, then only the standards matching the selected hole fit will be displayed in the **Shaft Fit** drop-down list and vice versa. However, if you select the **User Defined** option from the **Classification** drop-down list, you can select any standard from the **Hole Fit** and **Shaft Fit** drop-down lists. The **Stacked with line display** button is chosen to display the stacked tolerance with a line. You can also display the tolerance as stacked without a line using the **Stacked without line display** button. If you choose the **Linear display** button, the tolerance will be displayed in the linear form. The dimension along with the hole fit and shaft fit is shown in Figure-11.

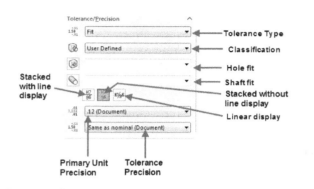

*Figure-10. The **Tolerance/Precision** rollout with the **Fit** option selected in the **Tolerance Type** drop-down list*

Figure-11. Hole fit and shaft fit

Fit with tolerance

The **Fit with tolerance** option in the **Tolerance Type** drop-down list is used to display tolerance along with the hole fit and shaft fit in a dimension. To apply fit with tolerance, select a dimension, and then select the **Fit with tolerance** option from the **Tolerance Type** drop-down list. Select the type of fit from the **Classification** drop-down list. Next, select the fit standard from the **Hole Fit** or the **Shaft Fit** drop-down list. Tolerance will be displayed with the fit standard only if you select a fit system from the **Hole Fit** or **Shaft Fit** drop-down list. Tolerance will be displayed along with the fit standard in the drawing area. In SolidWorks, tolerance is calculated automatically, depending on the type and standard of the fit selected. The **Show parentheses** check box can be selected to show tolerance in parentheses. The dimension along with the fit and tolerance is shown in Figure-12.

Figure-12. Dimensioning along with the fit and tolerance

Fit (tolerance only)

The **Fit (tolerance only)** option in the **Tolerance Type** drop-down list is used to display the tolerance in a dimension based on the hole fit or shaft fit.

None

The **None** option in the **Tolerance Type** drop-down list is used to display the dimensional value without any tolerance.

Unit Precision

The **Unit Precision** drop-down is used to specify the precision of the number of places after the decimal for dimensions. By default, the selected precision is two places after the decimal.

Tolerance Precision

The **Tolerance Precision** drop-down is used to specify the precision of the number of places after the decimal for tolerance. By default, the selected precision is two places after the decimal. This drop-down list will not be available, if the **None** option is selected in the **Tolerance Type** drop-down list.

Dimension Text Rollout

The **Dimension Text** rollout, as shown in Figure-13, is used to add text and symbols to dimension. The text box in this rollout is used to add text to dimension. The **<DIM>** text displayed in the text box symbolizes the dimensional value. You can add text before or after the dimension value. There are two text boxes in SolidWorks to write text above and below the dimension line. There are four buttons at the left of these text boxes. Choose the **Add Parentheses** button to enclose the dimension text in parentheses. Choose the **Inspection Dimension** button to enclose the dimension text in an obround shape and this dimension will be checked during inspection. Choose the **Center Dimension** button ⟨·x·⟩ to place the dimension at the center of the dimension line. If you need to place the text at a distance from the dimension line, choose the **Offset Text** button ⟨⟩ and drag the dimension to the required location. Note that you can also choose all four buttons for a specific dimension.

Figure-13. Dimension Text rollout

This rollout also provides buttons to modify text justification and add symbols such as diameter, degree, plus/minus, centerline, and so on to the dimension text. You can add more symbols by choosing the **More Symbols** button from the **Dimension Text** rollout and click on the **More Symbols** button from the flyout displayed. On choosing this button, the **Symbol Library** dialog box will be displayed, as shown in Figure-14.

Figure-14. Symbol Library dialog box

Dual Dimension Rollout

You need to select the check box in the **Dual Dimension** rollout to enable the options in this rollout, refer to Figure-15. The options in this rollout are used to display the alternative dimension value. Note that the alternative dimension value is displayed in square brackets, as shown in Figure-16. The options in this rollout are similar to those discussed in the earlier sections. Note that the alternative unit is set in the **Dual Dimension Length** cell in the **Document Property - Units** dialog box. To invoke this dialog box, choose **Tools > Options** from the SolidWorks menus; the **System Options** dialog box will be displayed. Choose the **Document Properties** tab; the name of this dialog box will be changed to the **Document Property - Drafting Standard** dialog box. In this dialog box, select the **Units** option from the area on the left to display the options for setting units. Note that on selecting the **Unit** option, the name of this dialog box will be changed to the **Document Property - Units** dialog box.

Figure-15. Dual Dimension rollout

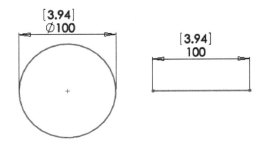

Figure-16. Entities with dual dimension

Sometimes, you may need to change the type of arrowheads or place the dimension at a distance from the entity because of space constraint. In SolidWorks, these actions can be performed by choosing the **Leaders** tab in the **Dimension PropertyManager**. The rollouts in this tab are discussed next.

Witness/Leader Display Rollout

The **Witness/Leader Display** rollout is used to specify the arrowhead style in dimensions, refer to Figure-17. The options in this rollout are discussed next.

*Figure-17. The rollouts in the **Leaders** tab*

Outside

The **Outside** button is used to display the arrows outside the extension line. To do so, select a dimension from the drawing area and choose the **Outside** button from the **Witness/Leader Display** rollout.

Inside

The **Inside** button is used to display the arrows inside the extension line. To do so, select a dimension from the drawing area and choose the **Inside** button. You can also click on the control point displayed on the arrowhead to reverse its direction.

Note that you can also click on the control point displayed on the arrowhead to reverse the direction of the arrowhead.

Smart

The **Smart** button is chosen by default and the arrows are displayed inside or outside the extension line, depending on the space available between the extension lines.

Directed Leader

This button is chosen to change the leader style of a dimension created on a surface by using the **DimXpert**.

Style

The **Style** drop-down list is used to select the style of the arrowhead. The unfilled triangular arrow is selected by default. You can select any arrowhead style for a particular dimension or dimension style. To change the arrowhead style, select a dimension from the drawing area and then the arrowhead style from the **Style** drop-down list.

Use document bend length

To change the length of a leader line after the bend, clear this check box and specify the length in the edit box below the check box. By default, the value specified in the **Document Properties - Detailing - Dimensions** dialog box will be displayed in the edit box given below this check box.

Leader/Dimension Line Style Rollout

To enable the options in the **Leader/Dimension Line Style** rollout, you need to clear the **Use document display** check box in it. After clearing this check box, the **Leader Style** and **Leader Thickness** drop-down lists will be enabled. The **Leader Style** drop-down list is used to specify the leader style and the **Leader Thickness** drop-down list is used to specify the thickness of the leader. You can also select the **Custom size** option from the **Leader Thickness** drop-down list and specify the thickness of the leader in the spinner available below this drop-down list as per your requirement.

Extension Line Style Rollout

To enable the options in the **Extension Line Style** rollout, you need to clear the **Use document display** check box and **Same as leader style** check box in it. The options in this rollout are same as discussed for **Leader/Dimension Line Style** rollout. These options modify the display of extension lines in dimensions.

Custom Text Position Rollout

The options in this rollout are used to specify the position of the text on a dimension line. Select the check box in the **Custom Text Position** rollout to enable the options in this rollout. The options in this rollout are discussed next.

Solid Leader, Aligned Text

If you choose this button, the leader line will be placed parallel to the dimension line along with the text.

Broken Leader, Horizontal Text

On choosing this button, leader line will be placed parallel to the horizontal axis along with the text.

Broken Leader, Aligned Text

Choose this button to place the dimension line to the center of the text. In this case, the text and the leader line will be placed parallel to the dimension line.

In SolidWorks, you can change the units and font of the dimension text by choosing the **Other** tab. The rollouts in this tab are discussed next.

Override Units Rollout

If you need to change the existing units of the dimension, select the check box in this rollout to expand it and select the units from the **Length Units** drop-down list.

Text Fonts Rollout

The font style set in the **Document Property - Detailing - Annotations Font** dialog box will be the default font style. To change the font style, clear the **Use document font** check box and change the font style by choosing the **Font** button in the **Text Fonts** rollout.

Options Rollout

If you select the **Read only** check box in this rollout, the dimensional value cannot be changed. If you select the **Driven** check box, the value will be the driven value.

Horizontal/Vertical Dimensioning between Points

As mentioned earlier, you can add a horizontal or vertical dimension between two points. To add any of these dimensions, choose the required button from the **Dimensions/Relations** toolbar or the **Smart Dimension** flyout in the **Sketch CommandManager**. Select the first point and then the second point. Next, specify a point to place the dimension; the **Modify** dialog box will be displayed. Enter a new dimension value in this dialog box and press ENTER.

Now, we will deal with real-world problems related to sketches.

Practical 1

Create the sketch as shown in Figure-18. Also dimension the sketch as per the figure.

Figure-18. Practical 1

Steps to be performed:

Below is the step by step procedure of creating the sketch shown in the Figure-18.

Starting Sketching Environment

- Start SolidWorks if not started already.
- Click on the **New** button from the Menu Bar. The **New SolidWorks Document** dialog box will display; refer to Figure-19.

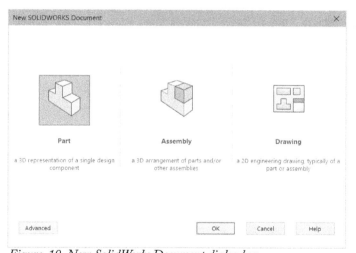

Figure-19. New SolidWorks Document dialog box

- Double-click on the **Part** button. The Part environment of SolidWorks will display as shown in Figure-20.

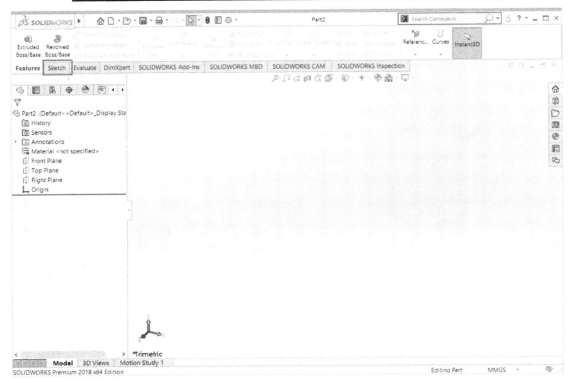

Figure-20. SolidWorks application window

- Click on the **Sketch** tab of the **Ribbon**; refer to Figure-20. The tools related to sketch will display in the **Ribbon**.

Starting Sketching

- Click on the **Sketch** button. Three default planes will be displayed.
- Select the **Front** plane from the viewport, refer to Figure-21. The viewport will become parallel to the view screen.

Figure-21. Selecting front plane

Creating Lines

- Click on the **Line** button from the **Ribbon**. The **Line** tool will become active and you are asked to select the start point.
- Click on the coordinate system and move the cursor towards right; refer to Figure-22.

Figure-22. Starting creation of line

- Enter **30** in the dimension box displayed below the line.
- Move the cursor vertically upwards and enter the value **40** in the dimension box.
- Move the cursor towards right and enter **20** in the dimension box. Refer to Figure-23.

Figure-23. Sketch after specifying 20 value

- Move the cursor upward and specify the value as **55** in the dimension box.
- Move the cursor towards left and specify the value as **30** in the dimension box.
- Move the cursor upward and specify the value as **40** in the dimension box.
- Move the cursor towards left and specify **20** in the dimension box.
- Move the cursor downward and click on the coordinate system to close the sketch.

The sketch after performing the above steps is displayed as shown in Figure-24.

Dimensioning the Sketch

- Click on the **Smart Dimension** button from the **Ribbon**. You will be asked to select entities.

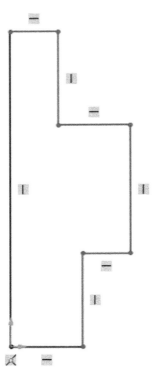

Figure-24. Completed Sketch

- Click on the bottom line joining with the coordinate system and move the cursor downwards; refer to Figure-25.
- Click at the appropriate distance to place the dimension. The **Modify** input box will display.
- Enter the dimension if you want to change. In this case, we will press **ENTER** to apply the default value.
- Click on the vertical line joining the end point of the bottom line selected earlier; refer to Figure-26.

Figure-25. Dimensioning bottom line

Figure-26. Vertical dimension to selected

- Place the dimension at proper distance from the line and press ENTER at the **Modify** input box.
- Similarly, dimension other entities of the sketch.

The sketch after applying all the dimensions will be displayed as shown in Figure-27.

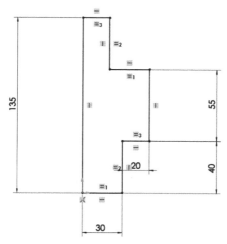

Figure-27. Final sketch after applying dimensions

Practical 2

In this practical, we will create the sketch as shown in Figure-28.

Figure-28. Practical 2

Steps to be performed:

Below is the step by step procedure of creating the sketch shown in the Figure-28.

Starting Sketching Environment

- Start SolidWorks if not started already.
- Click on the **New** button from the Menu Bar. The **New SolidWorks Document** dialog box will display.
- Double-click on the **Part** button. The Part environment of SolidWorks will display.
- Click on the **Sketch** tab of the **Ribbon**. The tools related to sketch will display in the **Ribbon**.

Starting Sketching

- Click on the **Sketch** button from the **Ribbon**. Three default planes will be displayed.
- Select the **Front** plane from the viewport. The viewport will become parallel to the view screen.

Creating Lines

- Click on the **Line** button from the **Ribbon**. You are asked to specify the start point of the line.
- Click on the coordinate system and enter **25** in the dimension box.
- Move the cursor down perpendicular to the previous line and enter **6** in the dimension box.
- Move the cursor to left and enter **12** in the dimension box.
- Move the cursor downwards and enter **50** in the dimension box. Till this point, our sketch should display like Figure-29.

Figure-29. Sketch after creating lines

Creating Arcs

- Click on the down arrow of **Arcs** drop-down and select **Tangent Arc** button from the list. You asked to click on an end point.
- Click on the end point of vertical line recently created and move the cursor downwards and then towards right until you get the preview as shown in Figure-30.
- When you get the preview like the Figure-30, click to create the arc. Don't worry about exact dimension but you need to make it a quarter of circle.

Figure-30. Arc creation

Creating Fillet

- Click on the **Sketch Fillet** tool from the **Ribbon** ⌐. The **Sketch Fillet PropertyManager** will display as shown in Figure-31.

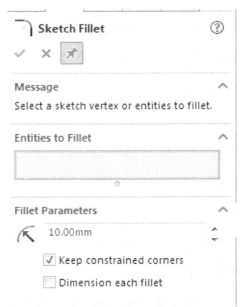

Figure-31. Sketch Fillet PropertyManager

- Click in the **Radius** spinner edit box in the **Parameters** rollout of the **PropertyManager** and enter the value as **3**.
- Select the lines as shown in Figure-32 for applying fillet. The fillet will be created between the two lines.
- Click on the **OK** button from the **PropertyManager**.

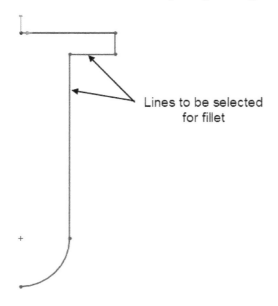

Lines to be selected
for fillet

Figure-32. Lines selection for fillet

Creating Mirror Copy

- Click on the **Centerline** tool from the **Line** drop-down. The **Line PropertyManager** will display as shown in Figure-33.

- Click on the coordinate system (start point of the sketch line) and then the end point of the arc. A center line will be created.
- Press **ESC** from **Keyboard** and exit the tool.
- Select the **Mirror Entities** tool from the **Ribbon** and select all the entities we have sketched except center line; refer to Figure-34.

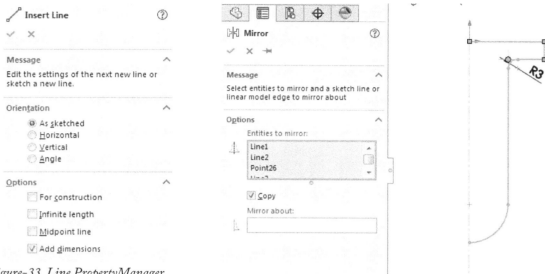

Figure-33. Line PropertyManager

Figure-34. Entities selected for creating mirror

- Click in the **Mirror about** box and select the center line from the sketch. Preview of mirror will be displayed; refer to Figure-35.

Figure-35. Mirror preview

- Click on the **OK** button from the **PropertyManager** to create the mirror.

Creating Circle and lines to complete

- Click on the **Circle** tool from the **Circle** drop-down in the **Ribbon**. The **Line PropertyManager** will be displayed; refer to Figure-36.

Figure-36. Circle PropertyManager

- Click on the **Diameter dimensions** check box from the **PropertyManager**.
- Click at the center point of the bottom arc and drag the cursor; refer to Figure-37.

Figure-37. Circle creation

- Enter the dimension as **13** in the input box.
- Click on the Tick mark from the **PropertyManager** to close it.
- Click on the **Line** tool from the **Ribbon** and click at any point on the center line to specify start point of the line; refer to Figure-38.

Figure-38. Point selected on centerline

- Move the cursor horizontally towards right and specify the value as 9.
- Move the cursor vertically downwards and click when the cursor is on arc.
- Press ESC to exit the tool.
- Mirror both the lines as we did earlier. The sketch should display as shown in Figure-39.

Figure-39. Sketch after all sketching operations

Dimensioning Sketch

- Click on the **Smart Dimension** tool from the Ribbon and select the arc. Dimension will get attached to cursor.

- Place the dimension at proper spacing. Press **ENTER** at the **Modify** input box.
- Click on the circle and place the dimension at proper place. Press **ENTER** at the **Modify** input box.
- Click on the two lines as shown in Figure-40.

Figure-40. Lines to be selected for dimensioning

- Click to place the dimension at its proper place. In the **Modify** input box, enter the value as **40**.

In the same way, dimension all the entities in the sketch until it is fully defined. The final sketch after dimensioning will be displayed as shown in Figure-41.

Figure-41. Final sketch

Practical 3

Create the sketch as shown in Figure-42. Also, dimension the sketch as per the figure.

Figure-42. Practical 3

Steps to be performed:

Below is the step by step procedure of creating the sketch shown in the Figure-42.

Starting Sketching Environment

- Start SolidWorks if not started already.
- Click on the **New** button from the Menu Bar. The **New SolidWorks Document** dialog box will display.
- Double-click on the **Part** button. The **Part** environment of SolidWorks will display.
- Click on the **Sketch** tab of the **Ribbon**. The tools related to sketch will display in the **Ribbon**.

Starting Sketching

- Click on the **Sketch** button from the **Ribbon**. Three default planes will be displayed.
- Select the **Top** plane from the viewport. The viewport will become parallel to the view screen.

Creating Circles

- Click on the **Circle** button from the **Circle** drop-down in the **Ribbon**. You are asked to specify the center point of the circle.
- Select the **Diameter dimensions** check box and **Add dimensions** from the **PropertyManager**.
- Click at the center of coordinate system to specify center of the circle.
- Enter the diameter as **1.125** in the edit box displayed.
- Again, click on the center of the coordinate system and specify the diameter value as **1.75** in the edit box.
- Click at the top left of the circles created to specify the center point of other circle; refer to Figure-43.

- Specify the diameter value as **0.75**.

Figure-43. Center position for circle

- Click at the center of newly created circle and specify the diameter value as **1.625**.
- Clear the **Diameter dimensions** check box from the PropertyManager and click on the center of coordinate system to specify center of a circle.
- Specify the radius of circle as **1.375** in the edit box.

Creating Slots

- Click on the **Centerpoint Arc Slot** button from the **Slot** drop-down. You are asked to specify the center point for construction circle of arc slot.
- Select the **Add dimensions** check box from the **Slot PropertyManager**.
- Click at the center of the coordinate system. Specify the radius of construction circle as **2.312**. You are asked to specify the starting point of the slot arc.
- One by one click at the two positions displayed in Figure-44. Move the cursor and you are asked to specify the width of the slot.
- Click in the screen when the reading of width is approximately **0.9**; refer to Figure-45.

Figure-44. Positions selected sor slot arc

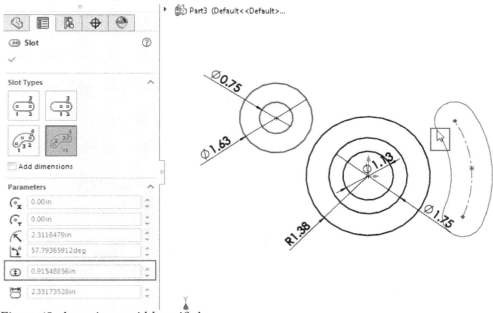

Figure-45. Approximate width specified

- Press the **Esc** key from the Keyboard and change the dimensions as shown in Figure-46.

Figure-46. Dimensioned arc slot

- Similarly, create other arc slot at the same location but with the corner radius of **0.875**; refer to Figure-47.

Figure-47. Sketch after creating second slot

- Click on the **Straight Slot** button from the **Slot** drop-down and create the slots as shown in Figure-48.

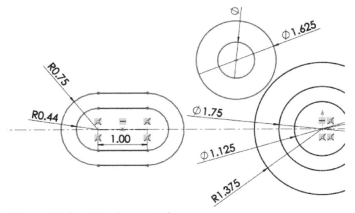

Figure-48. Straight slots created

Dimensioning not-dimensioned entities

- Click on the **Smart Dimension** tool and click on the center of circle with diameter **1.625**.
- Next, click on the center of coordinate system. Move the cursor vertically upward and place the dimension; refer to Figure-49.

Figure-49. Placing dimension

- Enter the distance value as **0.75**.
- Similarly, apply other dimensions; refer to Figure-50.

Figure-50. Sketch after applying dimensions

Creating Arcs

- Click on the **3 Point Arc** tool from the **Arc** drop-down in the **Ribbon**; refer to Figure-51. The **Arc PropertyManager** will be displayed; refer to Figure-52.

Figure-51. Arc drop-down

Figure-52. Arc PropertyManager

- Click one by one at the locations displayed in the Figure-53. Click on the **OK** button from the **PropertyManager** and change the value of radius to **1.75**.
- Select the arc and connected circle by holding the CTRL key, and select the **Tangent** button from the **Properties PropertyManager**; refer to Figure-54.

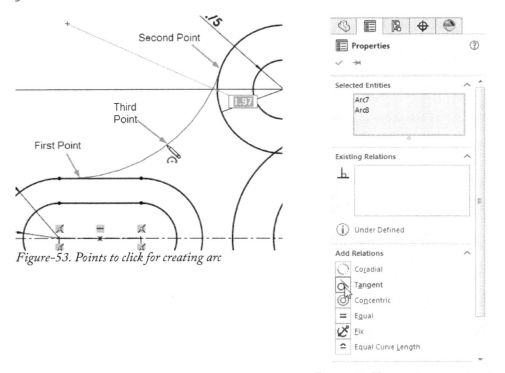

Figure-53. Points to click for creating arc

Figure-54. Tangent button to be selected

- Similarly, make the arc and slot tangent at the connecting point; refer to Figure-55.

Figure-55. Tangent arc created

- Click on the **Three Point Arc** tool again and similarly create the other arcs; refer to Figure-56.

Figure-56. Arcs to be created

Creating Line and Trimming

- Click on the **Line** tool from the **Line** drop-down. The **Insert Line PropertyManager** will be displayed.
- One by one click on the points shown in Figure-57.

Figure-57. Points selected for line

- Make the end points of the line tangent to slot and circle.
- Click on the **Trim Entities** tool from the **Trim Entities** drop-down and remove the extra sketched entities; refer to Figure-58.

Figure-58. Sketch after trimming extra entities

Practical 4

Create the sketch as shown in Figure-59. Also, dimension the sketch as per the figure.

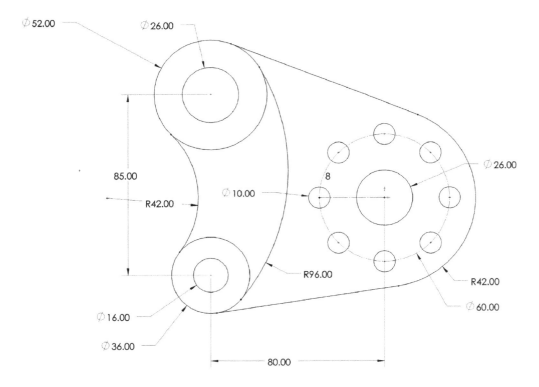

Figure-59. Practical 4 Sketching

Steps to be performed:

Below is the step by step procedure of creating the sketch shown in Figure-59.

Starting Sketching Environment

- Start SolidWorks if not started already.
- Click on the **New** button from the Menu Bar. The **New SolidWorks Document** dialog box will display.
- Double-click on the **Part** button. The **Part** environment of SolidWorks will display.
- Click on the **Sketch** tab of the **Ribbon**. The tools related to sketch will display in the **Ribbon**.

Starting Sketching

- Click on the **Sketch** button from the **Ribbon**. Three default planes will be displayed.
- Select the **Top** plane from the viewport. The viewport will become parallel to the view screen.

Creating Circles

- Click on the **Circle** button from the **Circle** drop-down in the **Ribbon** and create two circles at the origin of diameter **26** and **84**; refer to Figure-60.

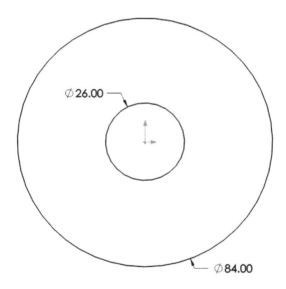

Figure-60. Circles created

- Create a centerline of length **80** towards left from the origin and then create circles as shown in Figure-61.

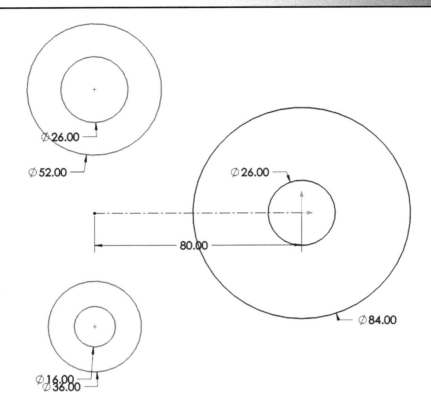

Figure-61. Circles created

- Select the center points of newly created circles and the end point of the centerline. The **Properties PropertyManager** will be displayed as shown in Figure-62.

Figure-62. Properties on selecting points

- Click on the **Vertical** button from the **Add Relations** rollout of the **PropertyManager**.

Creating Arcs

- Click on the **3 Point Arc** tool from the **Arcs** drop-down in the **Ribbon**. You will be asked to specify the starting point of the arc.
- Click on the points as shown in Figure-63 and specify the radius of the arc as **42** in the input box.

Figure-63. Points selected to create arc

- Select arc and top circle connected to it while holding the CTRL key. **Properties PropertyManager** will be displayed. Click on the **Tangent** button from **Add Relations** rollout; refer to Figure-64.
- Similarly, make the arc and bottom circle connected to it.
- Similarly, create an arc of radius 96 on the other side of these circles; refer to Figure-65. Make sure the arc is tangent to circles.

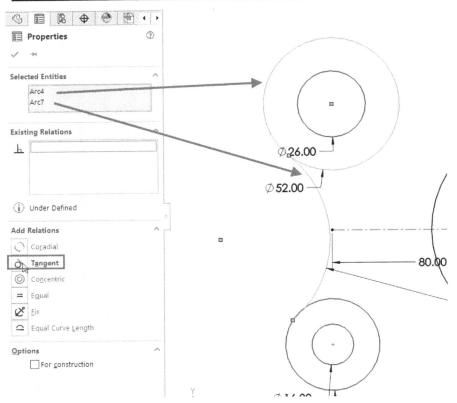

Figure-64. Making arc tangent to circle

Figure-65. Arc to be created

Creating Lines and Performing Trimming Operations

• Click on the **Line** tool from the **Line** drop-down in the **Ribbon** and create a line which is tangent to the circles as shown in Figure-66.

Note that after clicking on the first circle, you need to click on the other circle when a tangent mark is shown; refer to Figure-66.

Figure-66. Creating tangent line

- Similarly, create the line on other side of centerline; refer to Figure-67.

Figure-67. Tangent line created

- Click on the **Trim Entities** tool from the **Trim Entities** drop-down in the **Ribbon.** You will be asked to select the entities to be trimmed.
- Draw curve to trim the circle; refer to Figure-68. Click on the **OK** button from the **PropertyManager** to apply trimming.

Figure-68. Trimming circle

Creating Smaller Circle and Applying Pattern

• Click on the **Circle** tool from the **Ribbon** and create construction circle of diameter **60**; refer to Figure-69.

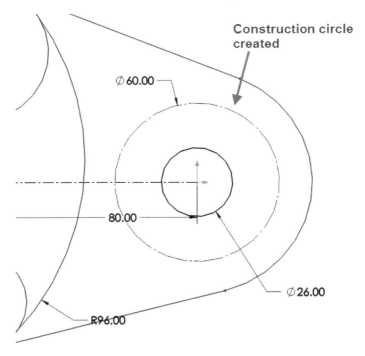

Figure-69. Creating construction circle

• Create a circle of diameter **10** at the intersection of centerline and construction circle; refer to Figure-70.

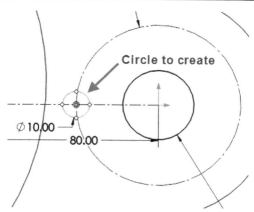

Figure-70. Small circle created for pattern

- Make sure the small circle created recently is selected and then click on the **Circular Sketch Pattern** tool from the **Sketch Pattern** drop-down in the **Ribbon**. Preview of pattern will be displayed along with **Circular Pattern PropertyManager**; refer to Figure-71.

Figure-71. Preview of circular pattern with PropertyManager

- Make sure **Equal spacing** check box is selected in the **Parameters** rollout of **PropertyManager** and **360 deg** is specified in the **Spacing** edit box. Enter **8** in the **Number of Instances** spinner of the **PropertyManager** and click on the **OK** button. The pattern of circles will be created.

Note that the center point of circular pattern was selected automatically on invoking the tool. If for some reason, the center is different from what is required then click in the **Center Point** selection box at the top in the **PropertyManager** and select the desired center point to modify the circular pattern.

- Now, dimension rest of the geometric entities and apply the constrains as required. The final sketch will displayed as shown in Figure-72.

Figure-72. Final sketch for Practical 4

Following are some sketches for practicing.

PRACTICE 1

In this practice session, we will create a sketch for the drawing given in Figure-73.

Figure-73. Practice 1

PRACTICE 2

In this practice session, we will create a sketch for the drawing given in Figure-74.

Figure-74. Practice 2

PRACTICE 3

In this practice session, we will create a sketch for the drawing given in Figure-75.

Figure-75. Practice 3

PRACTICE 4

In this practice session, we will create a sketch for the drawing given in Figure-76.

Figure-76. Practice 4

PRACTICE 5

In this practice session, we will create a sketch for the drawing given in Figure-77.

Figure-77. Practice 5

PRACTICE 6

In this practice session, we will create a sketch for the drawing given in Figure-78.

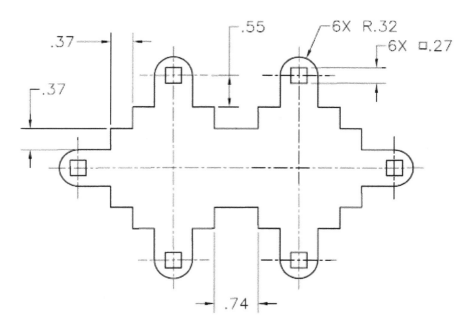

Figure-78. Practice 6

PRACTICE 7

In this practical, we will create a sketch for the drawing given in Figure-79.

Figure-79. Practice 7

To get more exercises, mail us at cadcamcaeworks@gmail.com

SELF ASSESSMENT

Q1. The .... **PropertyManager** is used to create, save, delete, and retrieve the dimension style in the Part environment.

Q2. On selecting a dimension, the .... **PropertyManager** is displayed.

Q3. The .... button is used to delete a dimension style.

Q4. The style file will be saved with the extension.............

a. .sldstl
b. .sldprt
c. .slddrw
d. .prt

Q5. Which of the following is a type of tolerance?

a. Limit
b. MIN
c. None
d. Both a and b

Q6. The inspection dimensions are enclosed in...........shaped box.

Q7. You can display dual dimensions in sketches of SolidWorks. (T/F)

Q8. The arrows are displayed inside or outside the extension line, depending on the space available between the extension lines by using the **Smart** button in **Leader Display** rollout. (T/F)

FOR STUDENT NOTES

FOR STUDENT NOTES

Chapter 4

3D Sketch and SolidModeling

Topics Covered

The major topics covered in this chapter are:

- *3D Sketching and Plane Selection.*
- *Convert Entities tool*
- *Extruded Boss/Base tool.*
- *Revolved Boss/Base tool.*
- *Swept Boss/Base tool.*
- *Creating Extra references for modeling.*
- *Lofted Boss/Base tool.*
- *Boundary Boss/Base tool.*
- *Hole Wizard*
- *Thread Tool*
- *Removing material using the above tools*

Extruded Revolved Swept Boss/Base Swept Cut Rib Wrap Referenc... Curves Instant3D
Boss/Base Boss/Base Lofted Boss/Bave Extruded Hole Wizard Revolved Lofted Cut Fillet Linear Pattern Draft Intersect
Boundary Boss/Base Cut Cut Boundary Cut Shell Mirror

Features Sketch Weldments Evaluate DimXpert SOLIDWORKS Add-Ins SOLIDWORKS MBD

3D SKETCHING

In previous chapter, you have worked on 2D drawings and have created them on Front plane. Now, we will move from 2D environment to explore the 3D environment. We will start with a 3D sketch and then we will create solid features in 3D environment.

A 3D sketch is the sketch which is not confined to one plane only. The 3D sketch can be in all the planes available in the viewport. To start with the 3D sketching, we are required to again open the **Sketch** tab in **Ribbon**. The steps to create a 3D sketch are given next.

- Click on the down arrow below the **Sketch** button. A list of tools will display.
- Click on the **3D Sketch** button from the list; refer to Figure-1.

Figure-1. 3D Sketch button

- Click on any of the sketching tool to create entities like, line, circle, rectangle and so on. (In our case, the **Line** tool is selected.)
- The cursor will start annotating the current sketching plane. Figure-2 shows the cursor which denotes that the current sketching plane is XY.

Figure-2. Current Sketching plane

- Click to specify the start point of the line in the XY plane.
- The point in 3D space where you click will become the current sketching plane.
- Enter the parameters for the line; refer to Figure-3.

Figure-3. Entering parameters of line

- Press **TAB** from the **Keyboard** to toggle between three standard planes; refer to Figure-4.

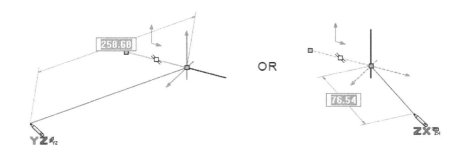

Figure-4. Toggling between planes

- Click or enter the parameters in the desired planes to create the sketch.
- You can also join two points in different planes by using the sketching tools; refer to Figure-5.

Figure-5. Connecting entities in 3D space

When we will be creating surfaces, we will use the 3D sketch again.

Convert Entities

The **Convert Entities** tool is used when you need the projection of any face, edge or sketch entity in another sketch. In this way, you can create the sketch entities from the projection of other features. The tool is available in the **Convert Entities** drop-down of the **Sketch** tab in the **Ribbon**. The procedure to use this tool is given next.

- Click on the **Convert Entities** tool from the **Convert Entities** drop-down in the **Ribbon**. The **Convert Entities PropertyManager** will be displayed; refer to Figure-6.

Figure-6. Convert Entities PropertyManager

- Select the face, edge or curve that you want to use in your current sketch; refer to Figure-7.

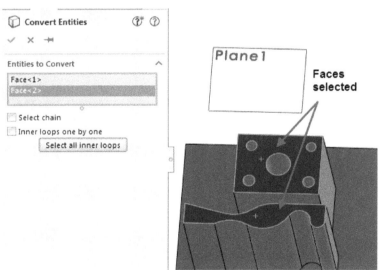

Figure-7. Selecting faces for converting entities

- Click on the **OK** button from the **PropertyManager**. The sketch entities will be generated; refer to Figure-8.

Figure-8. Entities converted

For any kind of analysis, simulation, assembly, or CAM; we need solid models. In SolidWorks, we are provided various tools to convert sketch into solid. Name of some of such tools are **Extruded Boss/Base** tool, **Revolved Boss/Base** tool, **Lofted Boss/Base** tool, and so on. These tools are explained next.

EXTRUDED BOSS/BASE TOOL

The **Extruded Boss/Base** tool is used to create a solid volume by adding height to the selected sketch. In other words, this tool adds material (by using the boundaries of sketch) in the direction perpendicular to the plane of sketch. In the term Boss/Base; the Base denotes the first feature and Boss denotes the feature created on any other feature. The steps to create extruded feature is given next.

- Click on the **Features** tab of the **Ribbon**. The tools related to solid modeling will display; refer to Figure-9.

Figure-9. Features Command Manager in Ribbon

- Click on the **Extruded Boss/Base** tool from the **Ribbon**. The **Extrude PropertyManager** will be displayed; refer to Figure-10.

Figure-10. Extrude PropertyManager

- Select a plane from the planes displayed in the viewport. The sketching environment will be displayed with the sketch tools activated.

- Create a closed sketch and then click on the **Exit Sketch** button from the viewport as shown in Figure-11. You can also select the **Exit Sketch** button from the **Ribbon**. The **Boss-Extrude PropertyManager** will display as shown in Figure-12.

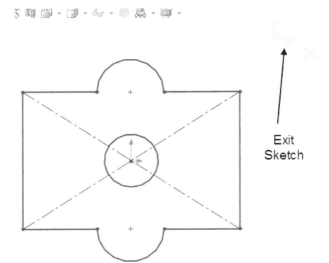

Figure-11. Sketch environment of extrude

Figure-12. Boss-Extrude PropertyManager

- Click on the **Starting reference** drop-down and select the desired option.

There are four options in this drop-down; **Sketch Plane, Surface/Face/ Plane, Vertex,** and **Offset**. The **Sketch Plane** is selected by default. Select this option if you want the extrusion to start from sketching plane. Select the **Surface/Face/Plane** option to start the extrusion from the selected surface/face/plane. Select the **Vertex** option to start extrusion from selected vertex. Select the **Offset** option if you want to start at specified distance from the sketching plane; refer to Figure-13.

Figure-13. Preview of offset option

• Click in the **Limiting reference type** drop-down and select the reference for end of extrusion.

There are six options in the drop-down; **Blind, Up To Vertex, Up To Surface, Offset From Surface, Up To Body,** and **Mid Plane**. If you have selected **Blind** or **Mid Plane** option, then you need to specify the distance value in the **Height of extrusion** spinner. If you have selected any of the other option then select the respective reference from the viewport. Figure-14 shows preview of extrusion by using the **Mid Plane** option. Note that if **Mid Plane** option is selected then the Direction 2 rollout will not display.

Figure-14. Mid-Plane extrusion

• Click in the **Direction of extrusion** selection box and select the reference if you do not want to extrude perpendicular to the sketching plane and want to extrude along selected axis/plane.
• Click in the edit box for extrusion height and enter the desired extrusion height or you can set the value by using spinner.
• Click on the **Draft On/Off** button to apply draft angle on the vertical faces of the model. On selecting this button, **1°** draft will be applied by default taking the sketching plane as reference. Select the **Draft outward** check box to apply draft angle outwards on the vertical faces of extrusion. Specify the draft angle in the **Draft Angle** spinner.

The parameters you specified above can also be applied to the opposite direction. To apply these parameters, select the **Direction 2** check box. The parameters for the opposite direction will display.

- Select the **Thin Feature** check box to create the thin walled extrusion. Enter the thickness in the **Thickness** edit box of the **Thin Feature** rollout. Figure-15 shows a thin featured extrusion. **Note that if open sketch is selected for extrude then this option gets selected automatically.**

Figure-15. Thin Feature extrusion

- If you want to close the start and end face of the extrusion then select the **Cap Ends** check box; refer to Figure-16.

Figure-16. Extrusion with cap ends

- Once you have finished creating the feature, click on the **Detailed Preview** button from the **PropertyManager** to verify the feature; refer to Figure-17.

Figure-17. Detailed Preview button

• Select the **Highlight new or modified faces** check box if you want to highlight the new/modified faces only. Similarly, you can select the **Show only new or modified bodies** check box if you want to display only new or modified objects.

In SolidWorks 2017 onwards, you can also activate the **Extrude Boss/Base** tool by ALT key. To do so, press the ALT key from keyboard and click on the shaded sketch section in the Sketching environment. The **Extrude** button will be displayed; refer to Figure-18. Click on the **Extrude** button to display the **Extrude PropertyManager**. Rest of the procedure is same as discussed earlier.

Figure-18. Alternate method for extrude

REVOLVED BOSS/BASE TOOL

The **Revolved Boss/Base** tool is used to create a solid volume by revolving a sketch about selected axis. In other words, if you revolve a sketch about an axis then the volume that is swept by revolved sketch boundary is called revolved boss/base feature. The steps to create revolved boss/base feature are given next.

• Click on the **Revolved Boss/Base** tool. If you have not selected any existing sketch then the **Revolve PropertyManager** will be displayed as shown in Figure-19.

Figure-19. Revolve PropertyManager

• Select a plane if you want to create a new sketch or select an already created sketch. In our case, we are selecting an already created sketch.

- Select the region of the sketch that you want to revolve if you have multiple loops in sketch; refer to Figure-20. The updated **Revolve ProperyManager** will be displayed as shown in Figure-21.

Figure-20. Region selected for revolve

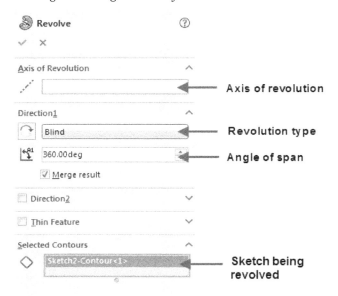

Figure-21. Updated Revolve Property Manager

- Click in the **Axis of Revolution** selection box to select the axis. Select the edge, line, or center line about which you want to revolve the sketch. Preview of the revolve feature will display; refer to Figure-22.

Figure-22. Preview of revolve feature

- Click on the **Revolve Type** drop-down and specify the revolution limiting reference. The options in this drop-down are same as discussed for **Extruded Boss/Base** tool.
- If you have selected **Blind** option in the **Revolve Type** drop-down then specify the degrees of revolution by using the **Angle** spinner.
- Click on the **Direction 2** check box to revolve in the direction opposite the earlier on. The options in the **Direction 2** rollout are same as discussed earlier.
- You can also create thin feature by selecting the **Thin Feature** check box. **Note that if you select an open sketch then this option is automatically selected.**
- Click in the **Selected Contours** box to add more sketches for revolution and select the sketches you want to revolve.

Figure-23 shows a sketch, axis of revolution, and resulting revolve feature preview.

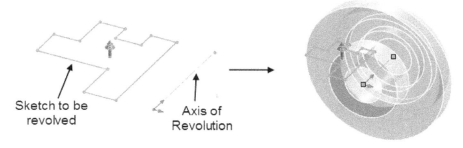

Figure-23. Revolved feature

SWEPT BOSS/BASE TOOL

The **Swept Boss/Base** tool is used to create a solid volume by moving a sketch along the selected path. In other words, if you move a sketch along a path then the volume that is covered by moving sketch boundary is called swept boss/base feature. Note that to use this tool, you must have a sketch section and a path, then only the tool will be active. The steps to create swept boss/base feature are given next.

- Click on the **Swept Boss/Base** tool. The **Sweep PropertyManager** will display as shown in Figure-24.

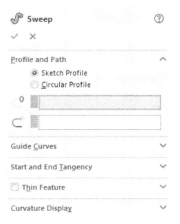

Figure-24. Sweep PropertyManager

- By default, the **Sketch Profile** radio button is selected and you are asked to select a sketch section (profile).
- Select the close section from viewport that you want to sweep.
- On selecting the section, the **Path** selection box becomes selected automatically and you are asked to select a path.
- Select the curve that you want to use as path. Preview of the sweep feature will be displayed.
- If your path is extended on both sides of section then three buttons will be displayed below the **Path** selection box; refer to Figure-25. Select the **Direction 1**, **Bidirectional**, and **Direction 2** button as per the requirement. Preview of sweep with these buttons is shown in Figure-26.

Figure-25. Buttons for bidirectional sweep

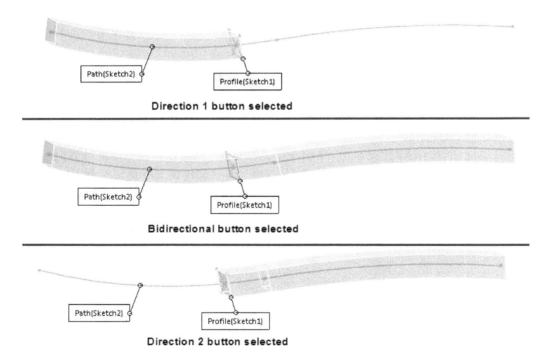

Figure-26. Sweep with directional options

- If your path is not perpendicular to the section plane then click on the **Options** rollout. The options in the rollout will display as shown in Figure-27.

Figure-27. Options rollout

- Click on the **Orientation/twist type** drop-down and select **Keep normal constant** option to create the sweep feature. Figure-28 shows a sweep feature created by this method.

Figure-28. Sweep with Follow Path and Keep Normal Constant options

Using Guide Curves

The **Follow Path** option is selected by default in the **Profile Orientation** drop-down and as a result the swept feature is created by making the sketch section exactly follow the curve's curvature. If you want to influence the shape of swept feature with the help of a guide curve then expand the **Guide Curves** rollout and select the guide curve/curves; refer to Figure-29. Select the **Follow path and First guide curve** option from the **Profile Twist** drop-down if you want to sweep feature to follow the path and first guide curve for its shape; refer to Figure-30. Similarly, you can select the **Follow First and Second Guide Curves** option from the drop-down to make the sweep feature follow both the guide curves for its shape.

Note that you need to have individual sketches for section, path, and guide curves although they can be on same plane.

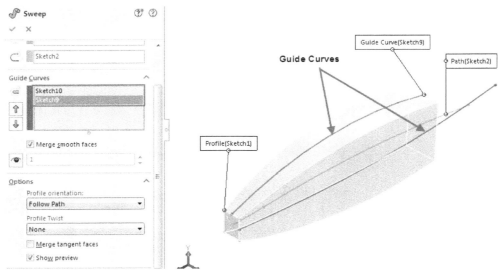

Figure-29. Sweep with two guide curves

Figure-30. Profile Twist drop-down

Applying Twist in Sweep Feature

You can twist the section while sweeping along the path to create drill bit type of shape of conduits. To do so, select the **Specify Twist Value** option from the **Profile Twist** drop-down in the **Options** rollout after selecting the section and path. The **Options** rollout will display as shown in Figure-31.

Figure-31. Options rollout with twist along path

Click on the **Twist Control** drop-down and select the desired unit to twist the section. In this case, we have selected **Revolutions** option. Specify the desired number of revolutions in the **Direction 1** spinner. Click on the flip button to reverse the twisting. Figure-32 shows the preview of twisting along the path while sweeping.

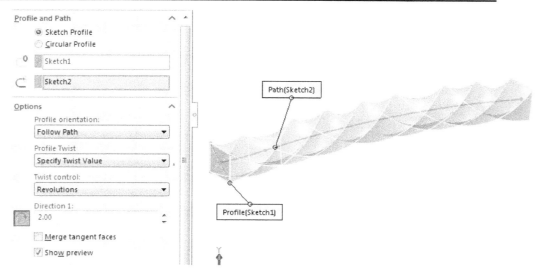

Figure-32. Preview of twist along path

You can also make a spring by using this option; refer to Figure-33.

Figure-33. Spring created by twist slong path

Circular Profile Sweep

This option is available in SolidWorks 2016 onwards version of software. Using this option, you can create round bar/rod. The procedure to use this option is given next.

- Select the **Circular Profile** radio button from the **Profile and Path** rollout of **PropertyManager**. You are asked to select a path.
- Select the path along which the circular sweep should be created. Preview of the circular sweep will be displayed; refer to Figure-34.

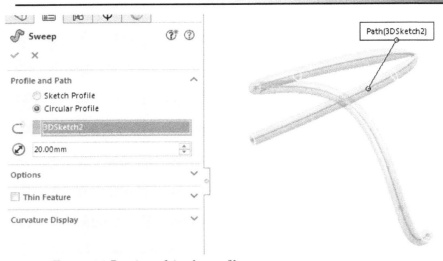

Figure-34. Preview of circular profile sweep

- Specify the desired diameter in the **Diameter** spinner and click on the **OK** button from the **PropertyManager**.

Before we move to **Lofted Boss/Base** tool and other feature creation tools. We need to understand the procedure to create reference planes, axis, coordinate system, and other references as they are important for advanced modeling.

REFERENCE GEOMETRY

There are various types of reference geometries that can be created in SolidWorks. All the tools to create these reference geometries are available in the **Reference Geometry** drop-down; refer to Figure-35. The tools available in the drop-down are:

- Plane
- Axis
- Coordinate System
- Point
- Center of Mass
- Bounding Box
- Mate Reference

Figure-35. Reference drop-down

These tools are discussed next.

Plane

The **Plane** tool is used to create reference planes. By default there are three planes available in SolidWorks: **Front**, **Top**, and **Right**. To create more planes follow the steps given next.

- Click on the **Plane** tool from the **Reference Geometry** drop-down. The **Plane PropertyManager** will display as shown in Figure-36.

Figure-36. Plane PropertyManager

- You can select maximum three references to create a plane. You can select plane/face, edge/axis/curve, or vertex/point. The ways in which you can create planes by using these references are discussed next.

Creating plane at a distance from plane/face

- To create a plane at a distance from a plane/face, select the plane/face. The updated **Plane PropertyManager** will display as shown in Figure-37.
- Specify the desired distance in the spinner.
- Click on the **OK** button to create the plane.

Figure-37. Updated Plane PropertyManager

Creating plane at an angle to plane/face

- Activate the **Plane PropertyManager** and select a plane/face to which you want to specify the angle.

- Click on the **At Angle** button to specify the angle.
- Click in the **Second Reference** box and select the edge or axis to which you want to make the plane coincident or select the two planar point through which you want the plane to pass. Figure-38 shows the plane create by both the methods discussed.

Figure-38. Plane creation at angle

Creating plane passing through points

- Activate the **Plane PropertyManager** and one by one click three points through which you want the plane to pass through. Figure-39 shows the plane passing through three points.

Figure-39. Plane passing through three points

Plane Parallel to Screen

This option is a new feature of SolidWorks 2016. The procedure to create plane parallel to screen is given next.

- Right-click on any face, edge, or vertex of the model. A shortcut menu will be displayed; refer to Figure-40.

Figure-40. Shortcut menu on right clicking on a face

• Click on the **Create Plane Parallel to Screen** option from the shortcut menu. A plane parallel to screen will be created; refer to Figure-41.

Figure-41. Plane parallel to screen

Axis

The **Axis** tool is used to create reference axes. An axis is useful in creating revolve features or to create planes at angle. To procedure to create axis by using the **Axis** tool is given next.

• Click on the **Axis** tool from the **Reference** drop-down. The **Axis PropertyManager** will display as shown in Figure-42.
• Select the desired button from the **PropertyManager**. The buttons in this **PropertyManager** are explained next.

Figure-42. Axis PropertyManager

One Line/Edge/Axis

Select this button if you want to create an axis coincident to the selected line/edge/axis. After selecting this button, click on the line/edge/axis. The axis will be created coincident to the selected line/edge/axis; refer to Figure-43.

Figure-43. Axis created on edge

Two Planes

Select the **Two Planes** button if you want to create axis at the intersection of the two selected planes/faces. After selecting this button, click on the two intersecting. The axis will be created at the intersection; refer to Figure-44.

Figure-44. Axis at intersection of planes

Two Points/Vertices

Select the **Two Points/Vertices** button if you want to create axis passing through the selected two points/vertices; refer to Figure-45.

Figure-45. Axis passing through two points or vertices

Cylindrical/Conical Face

Select the **Cylindrical/Conical Face** button and select a cylindrical/conical face. An axis passing through center of cylindrical/conical face will be created; refer to Figure-46.

Figure-46. Axis through cylinder and conical

Point and Face/Plane

Select the **Point and Face/Plane** button if you want to create an axis passing through the selected point and perpendicular to the selected face/plane.

Coordinate System

The **Coordinate System** tool is used to create reference coordinate system. The steps to create coordinate system are explained next.

• Click on the **Coordinate System** tool from the **Reference** drop-down. The **Coordinate System PropertyManager** will display as shown in Figure-47.

Figure-47. Coordinate System
PropertyManager

- Click on the point where you want to place the coordinate system.
- Click in the box for which you want to specify direction reference and select the reference like plane, axis and so on. Figure-48 shows a coordinate system created on the face.

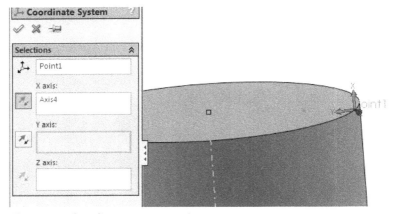

Figure-48. Coordinate system creation

Point

The **Point** tool is used to create reference points on the model. The steps to create points are given next.

- Click on the **Point** tool from the **Reference** drop-down. The **Point PropertyManager** will display as shown in Figure-49.
- Select the desired button to specify the type of point you want to create. In this case, we have selected **Center of Face** button.
- Select the reference (face of the model in this case). Preview of the point will display; refer to Figure-50.

Figure-50. Preview of point

Figure-49. Point PropertyManager

- Click on the **OK** button to create the point.

You can create array of points along a curve by selecting ![icon] button.

Center of Mass

The **Center of Mass** tool is used to display the center of mass of the model. The coordinates of center of mass are generally required in some calculations related to inertia of the objects. Identification of center of mass is also helpful in checking the stability of object in constraint free environment. To display the center of mass, click on the **Center of Mass** tool from the **Reference** drop-down and the center of mass will display in the viewport; refer to Figure-51.

Figure-51. Center of mass of cylinder

Bounding Box

The bounding box is used to create a minimum volume box around the selected part/assembly. The procedure to create bounding box is given next.

- Click on the **Bounding Box** tool from the **Reference Geometry** dropdown in the **Features CommandManager** in the **Ribbon**. The **Bounding Box PropertyManager** will be displayed; refer to Figure-52.

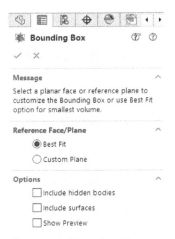

Figure-52. Bounding Box PropertyManager

- Select the desired check boxes to include hidden bodies and surfaces from the **Options** rollout.
- To check the preview, select the **Show Preview** check box.
- Select the **Custom Plane** radio button if the default selected plane is not appropriate for your requirement. After selecting this radio button, select the desired plane.
- Click on the **OK** button from the **PropertyManager** to create the bounding box.

LOFTED BOSS/BASE TOOL

The **Lofted Boss/Base** tool is used to create a solid volume joining two or more sketches created on different planes; refer to Figure-53. The procedure to create lofted features is given next.

Figure-53. Lofted feature example

- Click on the **Lofted Boss/Base** tool from the **Ribbon**. The **Loft PropertyManager** will display as shown in Figure-54.

Figure-54. Loft PropertyManager

- By default, **Profiles** selection box is active and you are asked to select sketches for lofted feature.
- Click one by one on the sketches created at different planes. Note that you need to select the sketches in the order by which they can be joined to each other successively. The preview of the lofted feature will display as shown in Figure-55.

Figure-55. Preview of the lofted feature

- Drag the green handle to align edges of the lofted feature. After aligning the edges, the above figure will be displayed as shown in Figure-56.

Figure-56. Lofted feature after aligning edges

- If you want to change the starting or end conditions of the loft feature then expand the **Start/End Constraints** rollout and select the desired option from the **Start Constraint** and **End Constraint** drop-down. Figure-57 shows the preview of model after changing the constraints.
- If you have guide curve/curves, then click in the Guide Curves box and select them to refine the shape of lofted feature; refer to Figure-58.

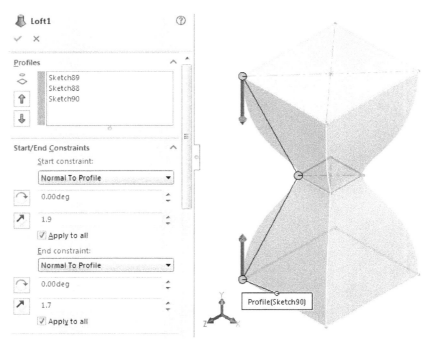

Figure-57. Preview after changing constraints

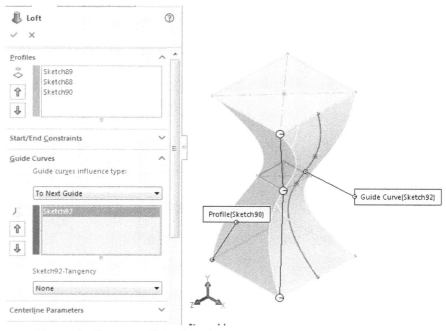

Figure-58. Preview of loft after selecting guide curves

- If you have a curve passing through center by which you want to control the shape of feature then expand the **Centerline Parameters** rollout and click in the selection box of this rollout. You are asked to select a center line or center curve
- Select the desired curve, the lofted feature will be modified accordingly.
- Next, click in the **Options** rollout to change the basic options of the lofted feature.
- Select the **Merge tangent faces** check box to merge the lofted feature with the adjoining tangent features.
- The **Close loft** check box is selected if you want to create a closed lofted feature. Note that to use this option, you need to clear the selection of guide curves by right-clicking in the **Guide Curves** box to use this feature. Figure-59 shows a closed lofted feature.

Figure-59. Preview of closed lofted feature

- Clear the **Merge Result** check box if you want to create it an individual object and don't want to join it with other solids in the model.

Note that the **Merge Result** check box will be available for each feature creation tool like **Extrude** and **Revolve**, after you have created the first solid/surface feature in the viewport. As discussed earlier, you can create a thin feature by using the options in the **Thin Feature** rollout.

BOUNDARY BOSS/BASE TOOL

The **Boundary Boss/Base** tool is used to create a solid volume by joining curves in different directions. On selecting this tool, the **Boundary PropertyManager** is displayed as shown in Figure-60. The tool works in the same way as the **Lofted Boss/Base** tool do. But, by using this tool, you can simultaneously analyze the surface of the feature being created and make it smoother or rougher as per requirement. Figure-61 shows the preview of the boundary boss/base feature with the mesh preview, Zebra stripes and Curvature combs.

Figure-60. Boundary PropertyManager

Mesh Preview　　　　　Zebra Stripes　　　　　Curvature Combs

Figure-61. Previews of Boundary Boss

The procedure to use the **Boundary Boss/Base** tool is given next.

• Create at least two sketch profiles for boundary base/boss feature. Note that the profiles should be intersecting each other at one point; refer to Figure-62.

• Click on the **Boundary Boss/Base** tool from the **Features CommandManager**. The **Boundary PropertyManager** will be displayed; refer to Figure-60.

• Select the base sketch. The sketch will be selected in **Direction 1** selection box.

• Click in **Direction 2** selection box in **PropertyManager** and then select the guide sketches. Preview of the boundary feature will be displayed; refer to Figure-63.

Figure-62. Sketches created for Boundary Base tool

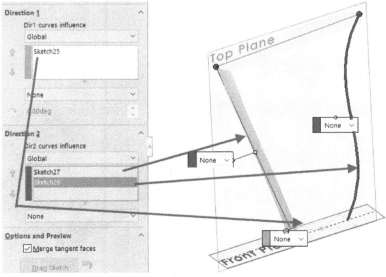

Figure-63. Preview of boundary feature

- Set the other parameters in **PropertyManager** as required and then click on the **OK** button to create the feature.

REMOVING MATERIAL FROM SOLID OBJECTS

Till this point, you have created base/boss features by using various feature creation tools. But in Engineering, machining an object means removing material from it. So, it is equally important to know the ways you can remove material from objects in SolidWorks. All the tools to remove material are available in the next column to the one discussed earlier; refer to Figure-64.

Figure-64. Tools to remove material

The tools except the **Hole Wizard** work in the same way as their respective Boss/Base tool. All these tools are explained one by one as follows:

Extruded Cut

The **Extrude Cut** tool is used to remove material by extruding the sketch. The steps to use this tool are given next.

- Click on the **Extruded Cut** tool. The **Extrude PropertyManager** will display.
- Select the face from which you want to start removing the material. The sketch environment will activate.
- Click the sketch using which you want to remove the material.
- Click on the **Exit Sketch** button from the **Ribbon**. The preview of cut feature will display; refer to Figure-65.

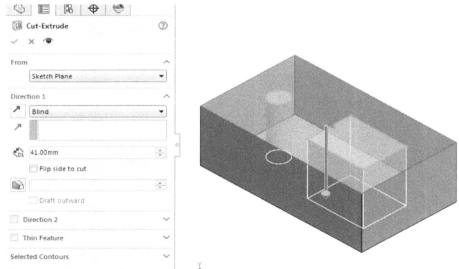

Figure-65. Preview of extrude cut feature

• Specify the depth of material removal and other parameters in the **PropertyManager** as discussed for **Extrude Boss/Base** tool and then Click on the **OK** button.

Revolved Cut

The **Revolve Cut** tool is used to remove material by revolving the sketch. The steps to use this tool are given next.

• Click on the **Revolved Cut** tool. The **Revolve PropertyManager** will display.
• Select the face from which you want to start removing the material. The sketch environment will be activated.
• Create the sketch of cut feature and a center line then click on the **Exit Sketch** button. Preview of revolved cut feature will display; refer to Figure-66.

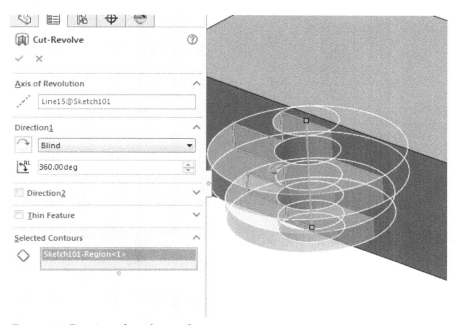

Figure-66. Preview of revolve cut feature

• Specify the parameters as discussed for **Revolved Boss/Base** tool and click on **OK** button to remove material.

Swept Cut

The **Swept Cut** tool is used to remove material by sweeping a section along the specified path. The steps to use this tool are given next.

• Click on the **Swept Cut** tool. The **Cut-Sweep PropertyManager** will display; refer to Figure-67. By default, the **Sketch Profile** radio button is selected in the **Cut-Sweep PropertyManager** and you are asked to select a sketch section.

- Select a closed sketch section and then an open sketch for path.
- Click on the **OK** button from the **PropertyManager** to create the swept cut; refer to Figure-68.

Figure-67. Cut Sweep PropertyManager

Figure-68. Swept cut with sketch profile

Swept Cut with Circular Profile

- Select the **Circular Profile** radio button from the **Cut-Sweep PropertyManager**. You are asked to select a path for the swept cut.
- Select an open sketch for the path. Preview of the cut will be displayed; refer to Figure-69.

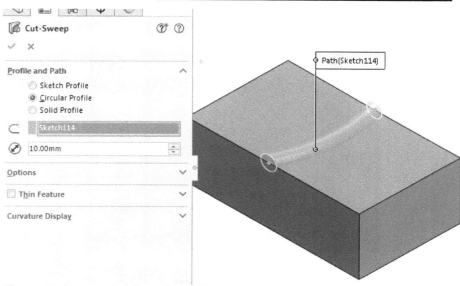

Figure-69. Swept cut with circular profile

- Specify the desired value for diameter of circular profile and click on the **OK** button from the **PropertyManager** to create the feature.

Swept Cut with Solid Profile

- Select the **Solid Profile** radio button from the **PropertyManager**. You are asked to select a solid revolve or cylindrical extrude feature as tool body. Note that the solid should be created by using only arc and line, and should not be merged with other bodies in model.
- Select the solid body and then select the path. Note that path should start from or within the tool body; refer to Figure-70.

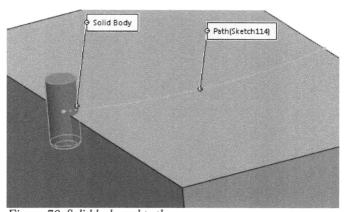

Figure-70. Solid body and path

- Click on the **OK** button from the **PropertyManager** to create the swept cut; refer to Figure-71. This option can be useful in checking the tool impression on a solid body before making any CAM programming.

The other tools like Lofted Cut and Boundary Cut work in the same way as their respective Boss/Base tool. I would be very happy if you practice those tools by yourself. If you get any doubt, please let me know at cadcamcaeworks@ gmail.com

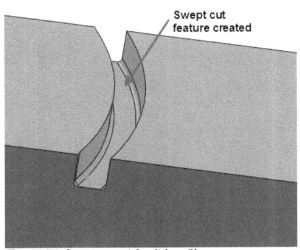

Figure-71. Swept cut with solid profile

Hole Wizard

The **Hole Wizard** tool is used to create holes that comply with the real machining tools. SolidWorks has library of standard holes and slots that can be created in the solid model. You can use this standard library or you can create a customized hole/slot by using the tool. The procedure to use this tool is given next.

- Click on the **Hole Wizard** tool from the **Hole Wizard** drop-down in the **Ribbon**. The **Hole Specification PropertyManager** will display as shown in Figure-72.

Figure-72. Hole Specification PropertyManager

- The **Favorite** rollout is used to store and reuse specific type of holes that you need again and again. Refer to Figure-73.

Figure-73. Favorite rollout

- Click on the desired type of hole from the **Hole Type** rollout. The parameters of the selected hole type will display in the **PropertyManager**.
- Select the desire hole standard and hole type from the **Standard** and **Type** drop-downs, respectively. The sizes of holes related to selected hole standard and type will display in the **Size** drop-down in the **Hole Specification** rollout.
- Click on the **Show custom sizing** check box, if you want to customize the hole.
- Enter the desired parameters and select the options as per your need from the other rollouts.
- Click on the **Add or Update Favorite** button from the **Favorite** rollout if you want to use this hole many times in the model.
- Now, click on the **Positions** tab to specify the position of the hole. The **PropertyManager** will display as shown in Figure-74.

Figure-74. Positions tab of Hole Specification PropertyManager

- Click on the face of the solid model where you want to place the hole. Click to specify the position of the hole. Note that later we can position it with dimensions.
- Click on the **OK** button to create the hole.
- On creating the hole, a node of it will be added in the **FeatureManager Design Tree**. Expand this node; refer to Figure-75.

Figure-75. Node of hole in FeatureManager Design Tree

- Right-click on the sketch for position of the hole and select the **Edit Feature** button from the tool box; refer to Figure-76.

Figure-76. Editing sketch

- The sketching environment will be displayed.
- Specify the position of point by dimensioning as you do for sketches. Note that here you can create multiple points by using the **Point** tool and the holes will be created at all the points you created. Refer to Figure-77.

Figure-77. Holes created at sketched points

Advanced Hole

The **Advanced Hole** tool is used to dynamically create holes in solids. This tool was first time introduced in SolidWorks 2017 version. The procedure to use this tool is given next.

- Click on the **Advanced Hole** tool from the **Hole Wizard** drop-down in the **Features CommandManager**. The **Advanced Hole PropertyManager** will be displayed along with the **Near Side** flyout; refer to Figure-78.
- Select the face to place the top of hole. Set the parameters for current element of hole in the **PropertyManager**; refer to Figure-79.
- Click on the **Insert Element Below Active Element** or **Insert Element Above Active Element** button from the **Near Side** flyout to insert more elements of hole in the near side. A new element will be added to the hole and in the **Near Side** flyout.
- Click on the down arrow next to newly added element and select the desired type of element from the drop-down displayed; refer to Figure-80.

Figure-78. Advanced Hole PropertyManager

Figure-79. Specifying parameters of hole element

Figure-80. Element types drop-down

- Set the parameters for newly added element in the **PropertyManager** as discussed earlier.
- Select the **Far Side** check box from the **Near and Far Side Faces** rollout in the **PropertyManager** if you want to specify the termination face of hole. The selection box for **Far Side** will be displayed in the rollout.
- Select the face for far side and select the desired elements from **Far Side** flyout displayed along the **PropertyManager**; refer to Figure-81.

Figure-81. Far Side selection box and flyout

- Set the parameters as discussed earlier.
- Expand the **Hole Callout** rollout to modify the callout. Select the **Customize callout** radio button. The box below the radio button will become edit able; refer to Figure-82.

Figure-82. Hole Callout rollout

- Select the desired callout string from the table to move it up or down and click on the **Move Up** button or **Move Down** button.
- To edit the parameter, double-click in the callout string and specify the desired parameter.

- Click on the **Position** tab in the **PropertyManager** and define the position of the hole.
- Click on the **OK** button to create the hole.

Thread

The **Thread** tool is used to cut helical thread on cylindrical faces. Using this tool, you can save custom threads in library. The procedure to use this tool is discussed next.

- Click on the **Thread** tool from the **Hole Wizard** drop-down in the **Features** tab of the **Ribbon**; refer to Figure-83. The **Thread PropertyManager** will be displayed; refer to Figure-84.

Figure-83. Thread tool

Figure-84. Thread PropertyManager

- Select the round edge of the cylindrical face of hole/boss feature. Preview of the thread will be displayed; refer to Figure-85.

Figure-85. Preview of thread

- Select the starting location of the thread after clicking in the **Start Location** selection box. 🗔▮▬▬▬▬ Note that this option is optional and you may not require it in your operation.
- Select the **Offset** check box if you want to start the thread at an offset distance from the starting location. After selecting this check box, an edit box becomes active below it where you can specify the value of offset; refer to Figure-86.

Figure-86. Thread at offset distance

- From the **End Condition** rollout, specify the depth of the thread. By default, the **Blind** option is selected in the drop-down and you need to specify the depth of thread in the edit box below it.
- Select the desired threading tool from the **Type** drop-down in the **Specification** rollout of the **PropertyManager**; refer to Figure-87.

Figure-87. Type drop-down for threading tools

- Select the thread size from the **Size** drop-down below the **Type** drop-down in the **Specification** rollout. You want to specify custom size of thread then click on the **Override Diameter** and **Override Pitch** buttons below the drop-down and specify the desired values; refer to Figure-88.

Figure-88. Custom size of thread

- Select the desired method from the **Thread Method** area of the rollout. There are two radio buttons available in this area; **Cut thread** and **Extrude thread**. Select the **Cut thread** radio button if you are creating thread inside a hole like in nut. Select the **Extrude thread** radio button if you want to create thread on a boss feature like bolt.
- You can cut/extrude the mirror image of current thread specifications by using the **Mirror Profile** check box. On selecting this check box, the **Mirror horizontally** and **Mirror vertically** radio buttons will become active. Select the **Mirror horizontally** radio button if you want to use horizontal mirror image of the current thread profile or select the **Mirror vertically** radio button if you want to use the vertical mirror image of the current thread profile; refer to Figure-89.

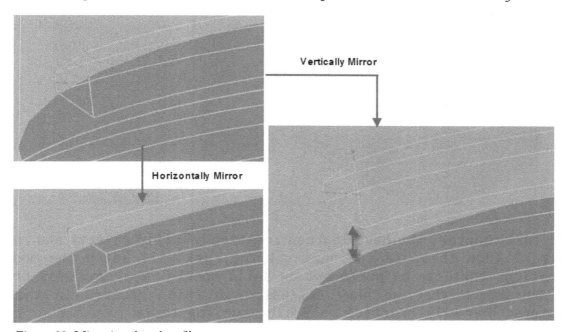

Figure-89. Mirroring thread profile

- Specify the thread angle using the **Rotation Angle** edit box.
- From the **Thread Options** rollout, you can select the thread as left handed or right handed.
- If you want to create a multiple start thread then select the **Multiple Start** check box from the **Thread Options** rollout and specify the desired number of starts in the spinner below the check box; refer to Figure-90. Note that the thread pitch value must permit multiple starts without causing crossing, self-intersecting threads.

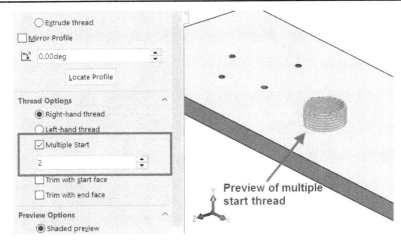

Figure-90. Multiple start thread

- To trim the thread at starting or end, select the **Trim with start face** or **Trim with end face** respectively from the **Thread Options** rollout.
- Check the preview of threads by using the desired radio button from the **Preview Options** rollout in the **PropertyManager**.
- Click on the **OK** button from the **PropertyManager** to create the thread; refer to Figure-91.

Figure-91. Threaded solids

Till this chapter, you have learned creation of Solid objects and removing material from them. In the next chapter, you will learn the modifying operations that can be performed on solid models. Now, you will practice the 3D sketch that we have discussed in this chapter. In the later chapters, you will practice on Solid Modeling tools.

PRACTICAL ON 3D SKETCH

Create a 3D sketch as shown in Figure-92.

Figure-92. 3D sketch for Practical

Starting Sketch

- Start SolidWorks by double-clicking on the SolidWorks 2018 icon from the desktop. (If not started yet.)
- Click on the **New** button from the **Quick Access Toolbar** or press CTRL+N from keyboard. The **New SOLIDWORKS Document** dialog box will be displayed.
- Double-click on **Part** icon from the dialog box. The Part environment of SolidWorks will be displayed.
- Click on **Sketch CommandManager** from the **Ribbon** and then click on the down arrow next to **Sketch** tool. A drop-down will be displayed.
- Click on the **3D Sketch** tool from the drop-down; refer to Figure-93.

Figure-93. 3D Sketch tool

Creating Lines in 3D

- Click on the **Line** tool from the **Sketch CommandManager** or press **L** from keyboard. The line tool will be active and coordinate system will be displayed as shown in Figure-94. Also, the name of active plane will be displayed below the cursor.

Figure-94. Coordinate system for 3D sketching

- Click on the coordinate system and press **TAB** key from keyboard till the YZ plane is active.
- Move the cursor along the negative Z axis and enter the length of line as **40** in the input box; refer to Figure-95.

Figure-95. Creating line along Z axis

- Move the cursor vertically upward along Y axis and enter the length as **35**; refer to Figure-96.

Figure-96. Creating line along Y axis

- Press TAB to activate ZX plane and create line of length **40** along negative Z axis; refer to Figure-97.

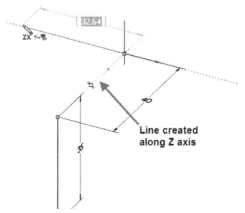

Figure-97. Line created along Z axis

- Move the cursor along X-axis and enter the length as 55 in the input box; refer to Figure-98.

Figure-98. Line created along X axis

- Repeat the same method to create rest of the lines; refer to Figure-99. Press ESC to exit the tool.

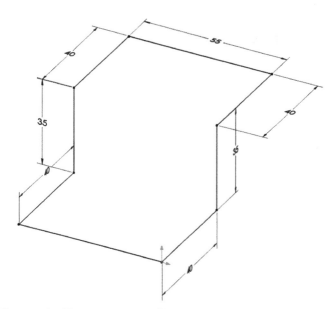

Figure-99. After creating outer lines

- Using the line tool join the inner points of the sketch created earlier; refer to Figure-100.

Figure-100. Creating lines to join points

Self Assessment

Q1. A sketch which exists in more than one plane is called ....

Q2. Using the...........key from keyboard, you can change the plane while creating the 3D sketch.

Q3. We can create a box using rectangle as sketch by using the ....tool.

Q4. We can create a cylinder using the rectangle as sketch by using the...............tool.

Q5. The.............tool is used to create a solid volume by moving a sketch along the selected path.

Q6. We can create sweep feature only at one side if the section lies in between the path. (T/F)

Q7. We can create both guide curves and path in same sketch for creating sweep feature. (T/F)

Q8. Using the **Profile Twist** options of the **Swept Boss/Base** tool, you can create springs. (T/F)

Q9. For Circular Profile Sweep feature, the profile of section is by default circle and we are required to specify the diameter of that. (T/F)

Q10. By default,............(hint: number) planes are available in SolidWorks.

Q11. How many number of maximum references can be provided for creating a reference plane?
a. 2
b. 3
c. 4
d. 5

Q12. After invoking the **Plane PropertyManager** if you select an existing plane then which of the following button is selected by default in the **PropertyManager**?

a. Parallel
b. Perpendicular
c. Coincident
d. Offset distance

Q13. Which of the following combination can be used to create plane at an angle?

a. An Edge and connected planar face
b. An Edge and intersecting plane
c. Two vertices and flat face
d. All of the above.

Q14. Which of the following cannot be selected as origin for creating coordinate system?

a. Vertices of solid
b. Edge of solid
c. Mid Point of an edge
d. Center line of revolve feature

Q15. Which of the following option is available for **Swept cut** tool but not for **Swept Boss/Base** tool?

a. **Circular Profile** radio button
b. **Solid Profile** radio button
c. **Guide Curves** rollout
d. **Mesh preview** check box

FOR STUDENT NOTES

Chapter 5

Solid Editing and Practical

Topics Covered

The major topics covered in this chapter are:

- *Fillet/Chamfer tool.*
- *Pattern tools.*
- *Rib tool.*
- *Draft tool.*
- *Shell tool.*
- *Wrap tool, Intersect tool, and Mirror tool.*
- *Equations*
- *Design Table*
- *Practical and Practice*

INTRODUCTION

In the previous chapter, we have learned to create solid models and remove material from them. In this chapter, we will learn to edit the solid models. Like the other tools, SolidWorks has packed all the editing tools into one column. These tools are available in the column next to the column intended for removing material; refer to Figure-1. The tools in this column are explained next.

Figure-1. Solid Editing tools

FILLET

The **Fillet** tool is used to apply radius at the edges. This tool works in the same way as the **Sketch Fillet** do. It is recommended that you apply the fillets after creating all the featured required in model, if possible because fillets can increase the processing time during modifications. The procedure to use this tool is given next.

Constant Size Fillet

- Click on the **Fillet** tool from the **Fillet** drop-down in the **Ribbon**. The **Fillet PropertyManager** will display as shown in Figure-2.

- Select the edge on which you want to apply the fillet. In place of selecting edge, you can select the two adjoining faces. The fillet will be created at the intersection of the two faces; refer to Figure-3.
- In the **Fillet Type** rollout **Constant size** button is selected.
- Specify desired radius for fillet in the **Radius** edit box in **Fillet Parameters** rollout of the **PropertyManager**.
- Select the **Full preview** radio button to check the preview of the fillet.
- You can select the desired profile fillet by using the **Profile** drop-down in **Fillet Parameters** rollout. Preview of fillet on selecting different profile options is given in Figure-4.

Figure-2. Fillet PropertyManager

Figure-3. Constant size fillet

Figure-4. Fillet with different profiles

- You can specify different radius for both sides of round with respect to selected edge by using the **Asymmetric** option from the drop-down in the **Fillet Parameters** rollout.
- You can set the setback parameters at the vertices by using the options in the **Setback Parameters** rollout; refer to Figure-5.

Figure-5. Setback Parameters rollout

- Click in the **Setback Vertices** selection box and click on the vertex for which you want to specify setback parameters. Fillet radius at the different joining edges will be displayed in the **Setback Distances** box.
- Select the desired direction from the **Setback Distances** box and specify the value of distance by using the **Distance** edit box in the **Setback Parameters** rollout.
- Click on the **OK** button from the **PropertyManager** to create the fillet.

Variable Radius Fillet

You can create fillet with varying radius by selecting the **Variable size** button. On selecting this button, the **Variable Radius Parameters** rollout is added in the **Fillet PropertyManager**; refer to Figure-6.

- Set the number of point for changing the radius by using the **Number of Instances** spinner in the rollout.
- Click in the **Attached Radii** box for start and end point in the viewport and specify the starting and ending radius.
- Click on the desired point in the preview of fillet and specify the desired radius; refer to Figure-7.
- You can change the profile for round by using the **Profile** drop-down in the rollout.

Figure-6. Variable Radius Parameters rollout

Figure-7. Creating variable radius fillet

Face fillet

- Select the **Face fillet** button from the **Fillet Type** rollout to create fillet at the joining edge of two faces.
- On selecting this radio button, the **Items to Fillet** rollout is added in the **Fillet PropertyManager**.
- Click in the first box and select the first face/faces.
- Next, click in the second box and select the second face/faces. Figure-8 shows preview of the fillet.

Full round fillet

Select the **Full round fillet** button to create a fillet where three faces meet each other. Using this option, you can create dome feature; refer to Figure-9.

Figure-8. Face fillet preview

Figure-9. Full round fillet preview

FilletXpert

The **FilletXpert** tool is used to apply different type of fillets in one single mode. The procedure to use this tool is discussed next.

- Click on the **FilletXpert** button from the **Fillet PropertyManager**. The **FilletXpert PropertyManager** will be displayed; refer to Figure-10.

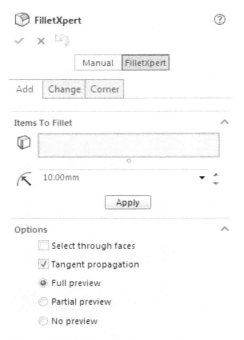

Figure-10. FilletXpert PropertyManager

- Select the face, edge, loop, or feature. Preview of the fillet will be displayed. If you have selected an edge then fillet will be applied to the edge. If you have select a face then all the edges of the face will be filleted. If you have selected a loop then all the edges connected to the loop will be filleted. If you have selected a feature then fillet will be applied to all the sharp edges of the feature.
- Set the desired value of fillet radius in the edit box below **Items To Fillet** selection box.
- Click on the **Apply** button to apply fillet and start with a new set of fillet.

CHAMFER

The **Chamfer** tool is used to bevel the sharp edges of the model. This tool works in the same way as the **Sketch Chamfer** do. The procedure to create chamfer by using this tool is given next.

- Click on the **Chamfer** tool from the **Fillet** drop-down. The **Chamfer PropertyManager** will display as shown in Figure-11.
- Select the edges from the model on which you want to apply chamfer.
- By default, the **Angle Distance** button is selected in the **Chamfer Type** rollout of the **PropertyManager**. Specify the angle and distance

parameters in the respective edit boxes in **Chamfer Parameters** rollout. Select the edges to create chamfers.

- If you want to specify distances for both sides of chamfer then select **Distance Distance** button and specify the parameters. Click on the edges to create chamfers.
- If you want to create chamfer at corners then select the **Vertex** button and select the corner of the model.

Figure-11. Chamfer PropertyManager

Figure-12 shows chamfer created at edges and vertexes.

Figure-12. Chamfers created on edges and vertex

There were two new options added to this tool in SolidWorks 2017 named; **Offset Face** and **Face Face**.

- Click on the **Offset Face** button from the **Chamfer Type** rollout if you want to create chamfer generated by offset of face adjacent to select edge; refer to Figure-13. Specify the chamfer distance in the edit box in **PropertyManager** and select the edge/face/feature to create chamfer.

Figure-13. Offset face chamfer

- Click on the **Face Face** button from the **Chamfer Type** rollout in the **PropertyManager** if you want to create chamfers on the edges formed by intersection of selected faces. Then, specify the chamfer distance in the edit box in **PropertyManager**. Click on the base face to be used for chamfer. Click in the **Face Set 2** selection box in the **PropertyManager** and then select the other face(s). Preview of chamfer will be displayed; refer to Figure-14.

Figure-14. Preview of chamfer

- After specifying desired parameters of chamfer, click on the **OK** button from **PropertyManager** to create the chamfer.

LINEAR PATTERN

The **Linear Pattern** tool is used to create linear pattern of solid features in the Modeling environment. This tool is similar to the sketch linear patterns. The procedure to create linear pattern is given next.

- Select all the features that you want to pattern from the **FeatureManager Design Tree** and select the **Linear Pattern** tool from the **Linear Pattern** drop-down. The **Linear Pattern PropertyManager** will display as shown in Figure-15.
- Select the direction reference for Direction 1 (like edge, face, plane, and so on) and specify the related parameters.

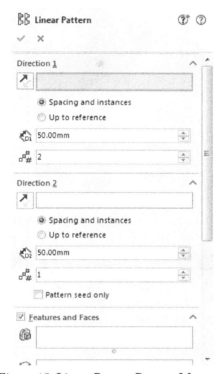

Figure-15. Linear Pattern PropertyManager

- Similarly, click in the **Direction 2** selection box in **PropertyManager** and specify the reference for Direction 2 and specify the related parameters; refer to Figure-16.

Figure-16. Preview of Linear pattern

- If you want to skip any instance in the pattern then expand the **Instances to Skip** rollout and select the pink dots in the preview to skip them; refer to Figure-17.

Figure-17. Skipping instances in pattern

- Click on the **OK** button from the **PropertyManager** to create the pattern.

CIRCULAR PATTERN

The **Circular Pattern** tool is used to create multiple instances of selected features around an axis. This tool is similar to the sketch circular pattern. The procedure to create circular pattern is given next.

- Select all the features that you want to pattern and select the **Circular Pattern** tool from the **Linear Pattern** drop-down. The **Circular Pattern PropertyManager** will be displayed as shown in Figure-18.
- Select the edge/axis/circular face(axis of circular face will be automatically selected) about which you want to create the pattern. Preview of the pattern will be displayed.
- Specify the required parameters like angle between two instances, number of instances, and so on for Direction 1 and Direction 2 as discussed under previous tool. The preview of pattern will display; refer to Figure-19.
- Click on the **OK** button to create the pattern.

Figure-18. Circular Pattern PropertyManager

Figure-19. Preview of circular pattern

CURVE DRIVEN PATTERN

The **Curve Driven Pattern** tool is used to create multiple instances of a solid features along the selected path. The procedure to create curve driven pattern is given next.

- Select all the features that you want to pattern and select the **Curve Driven Pattern** tool from the **Linear Pattern** drop-down. The **Curve Driven Pattern PropertyManager** will display as shown in Figure-20.

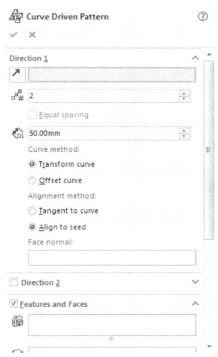

Figure-20. Curve Driven Pattern PropertyManager

- Select the curve along which you want to create the pattern.
- Specify the number of instances and other parameters. Preview of pattern will display; refer to Figure-21.
- Similarly, you can select a curve for direction 2 and specify the related parameters.

Figure-21. Preview of curve driven pattern

SKETCH DRIVEN PATTERN

The **Sketch Driven Pattern** tool is used to create multiple instances of a solid features as per the points specified in the selected sketch. The procedure to create sketch driven pattern is given next.

- Select all the features that you want to pattern and select the **Sketch Driven Pattern** tool from the **Linear Pattern** drop-down. The **Sketch Driven Pattern PropertyManager** will display as shown in Figure-22.
- Select the sketch as reference for pattern. Note that the sketch must have points created in it for referencing pattern instances. Preview of the pattern will be displayed; refer to Figure-23.
- Specify the other parameters as required and then click on the **OK** button from the **PropertyManager** to create the pattern.

Figure-22. Sketch Driven Pattern
PropertyManager

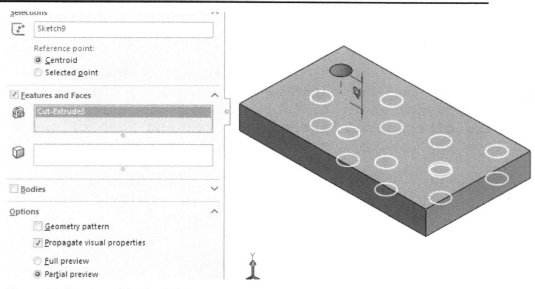

Figure-23. Preview of the sketch driven pattern

TABLE DRIVEN PATTERN

The **Table Driven Pattern** tool is used to create multiple instances of a features as per the coordinates specified in the table. Note that you must have a coordinate system created in the model for referencing the coordinates of the table. The procedure to create table driven pattern is given next.

- Select all the features that you want to pattern and then click on the **Table Driven Pattern** tool from the **Linear Pattern** drop-down in the **Ribbon**. The **Table Driven Pattern** dialog box will be displayed; refer to Figure-24.

Figure-24. Table Driven Pattern dialog box

- Click in the **Coordinate system** selection box and select the coordinate system. Note that you need to create a coordinate system at desired location before using this tool.
- Now, double-click in the cells of table and specify the desired coordinates for various instances of pattern. Preview of the pattern will be displayed; refer to Figure-25.
- Specify the other parameters as required and then click on the **OK** button from the dialog box to create the pattern.

Figure-25. Preview of the table driven pattern

- If you want to save the current pattern table then click on the **Save** button from the dialog box and save the file at the desired location.
- To use an already created pattern table, click on the **Browse** button. The **Open** dialog box will be displayed. Browse to the desired file double-click to use it as pattern table; refer to Figure-26.

Figure-26. Preview of table driven pattern

FILL PATTERN

The **Fill Pattern** tool is used to create multiple instances of a features by filling the selected close region. Note that you must have a closed loop sketch created in the model for referencing the region for fill pattern or you can select a flat face. The procedure to create fill pattern is given next.

- Select all the features that you want to pattern and then click on the **Fill Pattern** tool from the **Linear Pattern** drop-down in the **Ribbon**. The **Fill Pattern PropertyManager** will be displayed; refer to Figure-27.
- Select the sketch for filling with instances of pattern feature. Preview of the pattern will be displayed; refer to Figure-28.
- Specify the desired parameters like spacing between instances, space between boundary line and instances, and so on.
- Click on the **OK** button from the **PropertyManager** to create the fill pattern.

Figure-27. Fill Pattern Property-Manager

Figure-28. Preview of the fill pattern

Using Fill Pattern, you can create some predefined cuts in the solid model with the shapes of circle, square, diamond, and polygon. The procedure to use this option is given next.

- Without selecting any feature, click on the **Fill Pattern** tool from the **Linear Pattern** drop-down in the **Ribbon**. The **Fill Pattern PropertyManager** will be displayed.
- Select the **Create seed cut** radio button from the **Features and Faces** rollout of the **PropertyManager**. Buttons for various shapes will be displayed in the rollout with related parameter options; refer to Figure-29.

Figure-29. Options for creating seed cuts

- Select the desired shape button from the rollout and specify the parameters below it.
- Click on the face to apply the cuts. Preview of the fill pattern with seed cuts will be displayed; refer to Figure-30.
- Click on the **OK** button from the **PropertyManager** to create the feature.

Figure-30. Preview of fill pattern with seed cuts

MIRROR

The **Mirror** tool is used to create mirror copy of the features in the Modeling environment. This tool is similar to the sketch mirror. The procedure to create mirror is given next.

- Select all the features that you want to mirror.
- Click on the **Mirror** tool from the **Linear Pattern** drop-down in the **Ribbon**. The **Mirror PropertyManager** will display as shown in Figure-31.
- Select the plane or face about which you want to mirror the features. Preview of the mirror will be displayed; refer to Figure-32.
- Click on the **OK** button from the **PropertyManager** to create the feature.

Figure-31. Mirror PropertyManager

Figure-32. Preview of mirror

If you want to mirror a complete body with respect to mirror plane as in the above figure, then follow the steps given next.

- Click on the **Mirror** tool. The **Mirror PropertyManager** will display.
- Expand the **Bodies to Mirror** rollout and click on the bodies in the viewport that you want to mirror.
- Click in the **Mirror Face/Plane** box in the **Mirror Face/Plane** rollout of the **ProperyManager** and select the mirror plane. Preview of mirror will display; refer to Figure-33. Make sure that you clear the **Merge solids** check box before creating the mirror copy.
- Click on the **OK** button to create the mirror copy.

Figure-33. Preview of body mirror

RIB

The **Rib** tool is used to create support in the structures to increase their strength. You can find use of rib in various fixtures that are fastened to the wall or in the building columns to support objects. The procedure to create rib feature is given next.

- Click on the **Rib** tool from the **Ribbon**. You are asked to select a sketch or a sketching plane.

- Select the sketch if there is an existing one for rib or select the plane/face and create the sketch. Preview of the rib feature will display. Also, the options in the **PropertyManager** will be modified as per selection; refer to Figure-34.

Figure-34. Preview of rib feature

Note that the sketch should be created in such a way that its projection is within solid faces of the model; refer to Figure-35. For creating such sketch, you might need to create reference planes.

Figure-35. Sketch for rib

- Specify the thickness of rib feature and the draft angle in the corresponding spinners.
- You can flip the direction of rib by using the buttons for **Extrusion direction** in the **PropertyManager**. Note that the direction should be in such a way that the rib feature terminates by solid faces; refer to Figure-34.
- You can also change the thickness side by using the three buttons given for **Thickness**.
- Click on the **OK** button to create the rib feature.

DRAFT

The **Draft** tool is used to apply taper to the faces of a solid model. This tool is mainly useful when you are designing components for molding or casting. **Draft** tool applies taper on the faces and this taper allows easy and safe ejection of part from the dies. The procedure to use this tool is given next.

- Click on the **Draft** tool from the **Ribbon**. The **DraftXpert PropertyManager** will display; refer to Figure-36.

Using Draft PropertyManager

- Click on the **Manual** button at the top in the **PropertyManager**. The **Draft PropertyManager** will display; refer to Figure-37.

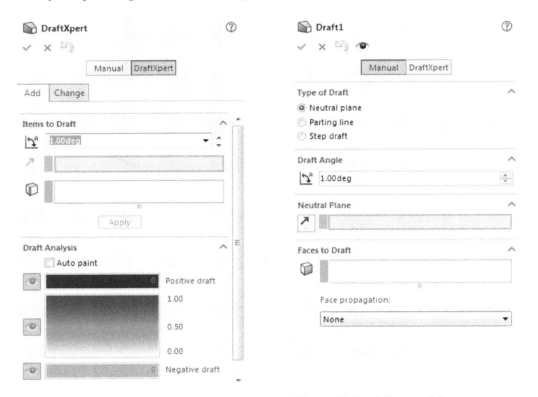

Figure-36. DraftXpert PropertyManager   *Figure-37. Draft PropertyManager*

- Make sure that the **Neutral Plane** radio button is selected and then select a face with respect to which you want to measure all the draft angles. You can flip the direction of draft by using the **Flip** button adjacent to the **Neutral Plane** box.
- On selecting the face, you are asked to select the walls on which you want to apply the draft.
- Change the draft angle by using the spinner in the **Draft Angle** rollout.
- Click on the **Detail Preview** button 👁 to check the preview of the draft. Click again on the button to exit the preview; refer to Figure-38.
- Click on **OK** button to create the draft.

We will learn about other options later in the chapter **Mold Tools.**

Figure-38. Draft preview

Using DraftXpert PropertyManager

- Click on the **Draft** tool from the **Ribbon** and then click on the **DraftXpert** button if not selected by default. The **DraftXpert PropertyManager** will be displayed.
- Select the face to be used as neutral plane. You will be asked to select the face(s) to which draft is being applied.
- Select the desired face and specify the draft angle in the angle spinner.
- Click on the **Apply** button from the **PropertyManager**. The draft will be created; refer to Figure-39.
- Click on the **OK** button from the **PropertyManager** to exit the tool.

Figure-39. Applying draft using DraftXpert

SHELL

The **Shell** tool is used to make a solid part hollow and remove one or more faces. The procedure to use this tool is given next.

- Click on the **Shell** tool from the **Ribbon**. The **Shell PropertyManager** will display; refer to Figure-40.
- Specify the desired thickness in the spinner.
- Click on the **Show preview** check box to display the preview.
- Select a face that you want to remove.
- You can flip the direction of shell by clicking on the **Shell outward** check box.
- Click on the **OK** button to create the feature. Figure-41 shows the preview of the shell feature and output of shell.

Figure-40. Shell PropertyManager

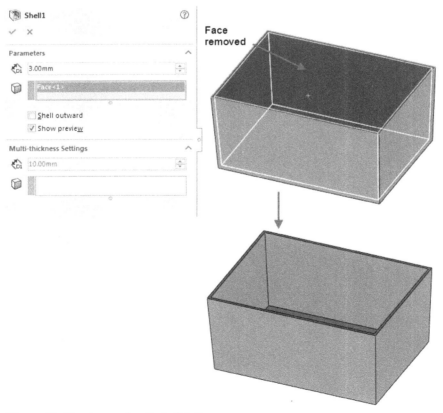

Figure-41. Preview and output of shell

WRAP

The **Wrap** tool is used to wrap text or any curve/curves on the cylindrical faces. Before using this tool, create a sketch that you want to wrap on a plane parallel to the wrapping face; refer to Figure-42. The steps to create wrap are given next.

Figure-42. Sketch for wrap

- Select the sketch and then click on the **Wrap** tool. The **Wrap PropertyManager** will display as shown in Figure-43.

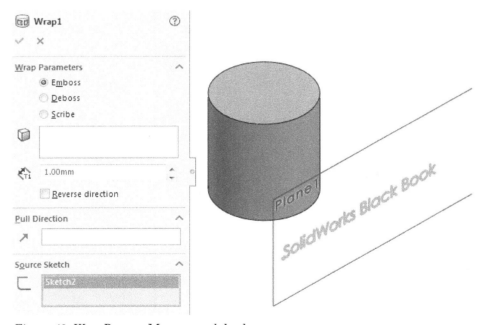

Figure-43. Wrap PropertyManager and sketch

- Select the desired radio button for embossing, engraving (debossing), or scribing.
- Select the cylindrical face on which you want to emboss or engrave the sketch. Preview will display.
- Set the thickness value in the spinner.
- Click **OK** button to create the feature. Figure-44 shows a sketch embossed on the cylindrical face.

Figure-44. Output of wrap

INTERSECT

The **Intersect** tool is used to separate the volume create by intersection of solids, surfaces or a combination of both. This tool can be very helpful for creating mold tools manually. The steps to use this tool are given next.

- Click on the **Intersect** tool from the **Ribbon**. The **Intersect PropertyManager** will display as shown in Figure-45.
- Select the surfaces/solids from which you want to extract the intersecting volume.
- There are two type of regions formed while intersecting solids; Intersecting regions and internal regions formed during intersection. There are two radio buttons in the **PropertyManager** to find out these regions; **Create intersecting regions** and **Create internal regions**. You can select the **Create both** radio button to include both regions in the list.
- Click on the **Intersect** button from the **PropertyManager**, the preview will be displayed in the viewport. Also, the region that you can exclude will be displayed in the **Regions to Exclude** rollout; refer to Figure-46.

Figure-45. Intersect PropertyManager

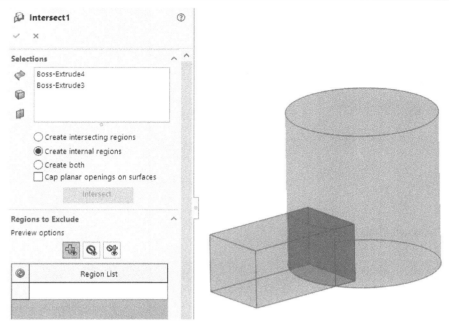

Figure-46. Preview of intersection

- Click on the check boxes of feature that you want to exclude from the **Regions to Exclude** rollout or select the portion of model from the viewport.
- Click on the **OK** button from create the intersection portion. Figure-47 shows the output of the intersection.

Figure-47. Output of intersection

EQUATIONS

The **Equations** tool is used to manipulate the model based on specified equations for various parameters of part. The procedure to use **Equation** tool is given next.

- Click on the **Equations** tool from the **Tools** menu; refer to Figure-48. The **Equations, Global Variables, and Dimensions** dialog box will be displayed; refer to Figure-49.

Figure-48. Equations tool

Figure-49. Equations, Global Variables, and Dimensions dialog box

- By default, the **Equation View** button is selected in the dialog box so, you can see the parameters related to equations like global variables, features suppressed/unsuppressed, and equations. Similarly, select the **Sketch Equation View** button to check equations in various sketches of part.
- Click on the **Dimension View** button from the dialog box as we will be writing equations in this mode. On selecting this button, all the dimensions of the model will be displayed in the dialog box; refer to Figure-50.

The options in this dialog box are divided into three nodes viz. Global Variables, Features, and Dimensions. Global variables are parameters which are user-defined and are used in equations. Like, you can create a global variable scale and use it to scale-up or scale-down the model. In the **Features** node, you can give the name of feature and set it suppressed/unsuppressed. In the **Dimensions** node, you can set the equations for various dimensions. The methods to use options in each node are discussed next.

Figure-50. Dimension View button selected in dialog box

Adding Global Variables

- Click in the **Add global variable** field in the **Global Variables** node of the dialog box and type desired name of the variable like scale in the format **"scale"** (note that "" are included). Press **TAB** from keyboard. You will be asked to specify the value of variable.
- Type the desired value of variable in the edit box and then press **ENTER**. The global variable will be created; refer to Figure-51.

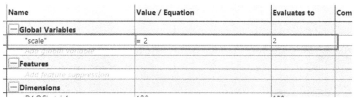

Figure-51. Global variable created

Suppressing/Un-suppressing Features

- Click in the **Add feature suppression** field of the **Features** node. You are asked to select the feature to be suppressed/unsuppressed.
- Select the feature from **FeatureManager Design Tree**. A flyout will be displayed in the dialog box with various options; refer to Figure-52.

Figure-52. Flyout displayed for features

- Select **suppress** or **unsuppress** option from the **Global Variables** menu in the flyout. Note that any global variable created by you earlier will also be displayed in this menu.

Setting Dimension Equations

- Click in the value area of the dimension that you want to solved by equation and remove its value.
- Type the equation for dimension in the field; refer to Figure-53. Press ENTER to apply the equation.
- Similarly, create other equations and then click on the **Rebuild** button ❽ from the dialog box to update the model; refer to Figure-54.
- Select the **Automatically Rebuild** check box if you want to model to be updated automatically. Click on the **OK** button to apply equations.

Figure-53. Equation typed for dimention

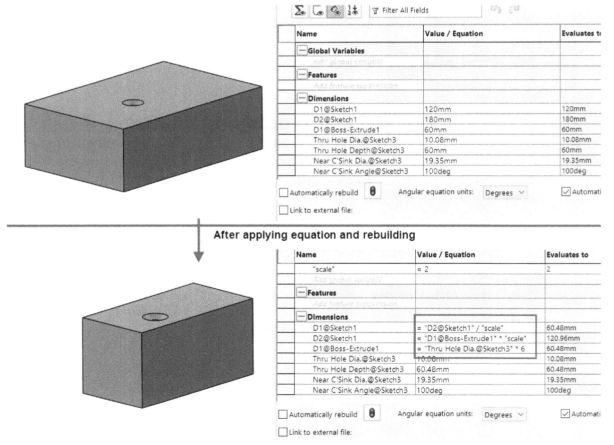

Figure-54. Applying dimensional equations

DESIGN TABLE

Design table is a very fast way to create multiple configurations of a part. Suppose you want to create multiple sizes of same shaped part like bolt, then you can use the Design table to create multiple parts while modeling the part only one time (Note that the part file is available in the resource kit). The procedure to use the **Design Table** tool is given next with the help of an example shown in Figure-55.

Figure-55. Wing Nut

- Create the model while taking the value of **D** as **10**; refer to Figure-56.

Figure-56. Wing nut created

- Create the equation of part as per the drawing given in Figure-55. The parameters to be specified for equation are given in Figure-57.

Name	Value / Equation	Evaluates to	Com
⊟Global Variables			
Add global variable			
⊟Features			
Add feature suppression			
⊟Dimensions			
D4@Sketch1	= 10	10mm	
D1@Sketch1	= "D4@Sketch1" * 2	20mm	
D2@Sketch1	= "D4@Sketch1"	10mm	
D3@Sketch1	= "D4@Sketch1" * 1.5	15mm	
D1@Revolve1	360deg	360deg	
D8@Thread1	0deg	0deg	
D7@Thread1	1mm	1mm	
D6@Thread1	= "D4@Sketch1"	10mm	
D4@Thread1	= "D4@Sketch1"	10mm	
D1@Thread1	0deg	0deg	
D1@Sketch7	= "D4@Sketch1" * 0.6	6mm	
D2@Sketch7	= "D4@Sketch1" * 0.4	4mm	
D3@Sketch7	= "D4@Sketch1" * 1.2	12mm	
D4@Sketch7	= "D4@Sketch1" * 1.4	14mm	
D1@Boss-Extrude2	= "D4@Sketch1" * 0.2	2mm	

Figure-57. Parameters for wing nut specified for equation

- Click on the **Design Table** tool from the **Insert -> Tables** menu; refer to Figure-58. The **Design Table PropertyManager** will be displayed; refer to Figure-59.

Figure-58. Design Table tool

Figure-59. Design Table PropertyManager

- Click on the **OK** button from the **PropertyManager**. The **Dimensions** selection box will be displayed; refer to Figure-60. Select the dimension which is used in equation to drive other dimensions and click on the **OK** button from the selection box. The design table excel sheet will be displayed; refer to Figure-61.

Figure-60. Dimensions selection box

Figure-61. Design table

- Under the first column specify the name of configuration like Wing Nut dia 12, Wing Nut dia 15 etc. In the second column specify the respective dimension value like 12, 15 etc. Refer to Figure-62.

Figure-62. Data entered in design table

- Click on the **OK** button displayed in the top right of drawing area (a tick mark). A notification will be displayed telling you that configurations have been generated through design table; refer to Figure-63.

Figure-63. Notification box

- Click on the **OK** button. The configurations will be created. Double-click on the desired configuration in **ConfigurationManager** to check the model; refer to Figure-64.

Figure-64. ConfigurationManager

APPLYING MATERIAL TO PART

Material is a very important properties of objects created in any CAD software. Most of the analysis and cost estimates are directly based on the material of part. The procedure to apply material to the part is given next.

- After creating model, click on the **Edit**->**Appearance**->**Material** tool from the Menu bar; refer to Figure-65. The **Material** dialog box will be displayed; refer to Figure-66.

Figure-65. Material tool

Figure-66. Material dialog box

- Select the desired material from the list and click on the **Apply** button to apply it on the current part.
- Click on the **Close** button to exit the dialog box.

MASS PROPERTIES OF PART

Click on the **Mass Properties** tool from the **Evaluate CommandManager** in the **Ribbon**. The mass properties of the part will be displayed; refer to Figure-67.

Figure-67. Mass Properties dialog box

Click on the **Close** button to close the dialog box.

Practical 1

Create the model(isometric view) as shown in Figure-68. The views of the model with dimensions are given in Figure-69.

Figure-68. Practical1 model

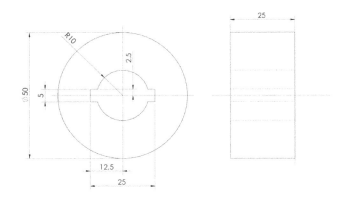

Figure-69. Views for Practical 1

Before we start working on the Practical, it is important to understand two terms; first angle projection and third angle projection. These are the standards of placing views in the engineering drawing. The views placed in the above figure are using third angle projection. In first angle projection, the top view of model is placed below the front view and right side view is placed at left of the front view. You will learn more about projection in chapter related to drafting.

Starting SolidWorks Modeling environment and creating Extrude feature

* Double-click on the SolidWorks icon from desktop if you have not started SolidWorks. Start a new part file.
* Click on the **Features CommandManager** in the **Ribbon** if not selected.

From the isometric view as well as from the other views, we can judge that this model can be easily created by extruding the sketch.

* Click on the **Extrude Boss/Base** tool. The **Extrude PropertyManager** will display.
* Select the **Top** plane from the **FeatureManager Design Tree** or from the viewport; refer to Figure-70.

Figure-70. Selecting Top plane

We will draw the sketch on top plane to get the isometric view as shown in Figure-68.

- Click on the **Normal To** button from the **View Orientation** drop-down in the **Heads-up View toolbar** or press **CTRL + 8** if the plane is not parallel to screen. Note that **8** key in shortcut is not from Numpad of keyboard.
- Click on the **Circle** tool from the **Ribbon**, select the **Diameter Dimensions** check box.
- Click at the coordinate system to place center of the circle.
- Drag the cursor and enter the value as **50** in the **Dimension box**.
- Again click at the coordinate system and draw circle of diameter **20**.
- Click on **OK** button from the **Circle PropertyManager**.
- Select the **Center Rectangle** tool from the **Rectangle** drop-down in the **Ribbon** and click at the center of the circles.
- Drag the cursor and specify the dimension of rectangle as **5** and **25** for height and width respectively. Click **OK** from the **PropertyManager**. The sketch after performing the above steps will display as shown in Figure-71.
- Select the **Trim Entities** tool from the **Ribbon** and trim the entities in such a way that the sketch is displayed as shown in Figure-72.

Figure-71. Sketch after creating circles and rectangle

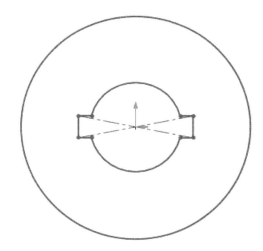

Figure-72. Sketch after trimming

- Click on the **Smart Dimension** tool and dimension the sketch. Refer to Figure-73.
- Click on the **Exit Sketch** button. Preview of the Extrude feature will display as shown in Figure-74.

Figure-73. Dimensioned sketch

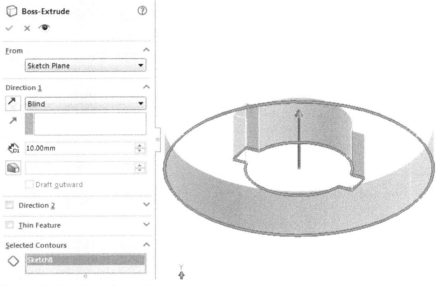

Figure-74. Preview of extrude

- Specify the height of extrusion as **25** and then click on the **OK** button from the **PropertyManager**. The model will be created as in Figure-68.

Practical 2

Create the model(isometric view) as shown in Figure-75. The dimensions and view is given in Figure-76.

Figure-75. Model for Practical2

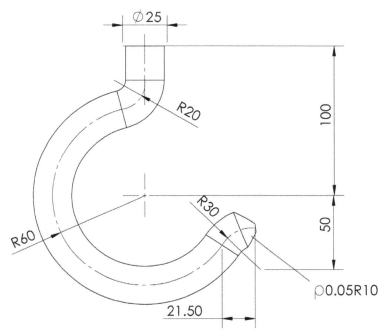

Figure-76. Practical2 drawing view

Start the SolidWorks if not started and open the modeling environment as explained in previous practical.

You can find out from the model that this hook can be easily created with the help of **Swept Boss/Base** tool. But before we use that tool, we must have sketches for path as well as section. The steps to create them are as follows:

Creating Sketches for Hook

- Start a new sketch on Front Plane.
- Click on the **Circle** tool and create the circle of diameter **120** taking coordinate system as center.
- Draw a straight line starting from top quadrant point and having length of approximately **40**.

- Draw a three point arc in the bottom area of the circle; refer to Figure-77. (It should look like the one given in figure. Accuracy is no required.)

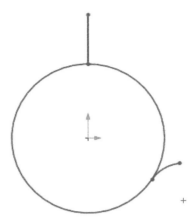

Figure-77. Sketch after creating three point arc

- Click on the **Trim Entities** tool and trim the portion between the straight line and the arc; refer to Figure-78.

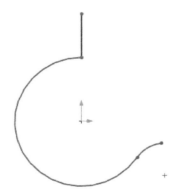

Figure-78. Sketch after trimming circle

- Click on the **Sketch Fillet** tool and apply fillet between the straight line and circle. Specify the fillet radius as **20**.
- Now, dimension the sketch as shown in Figure-79.

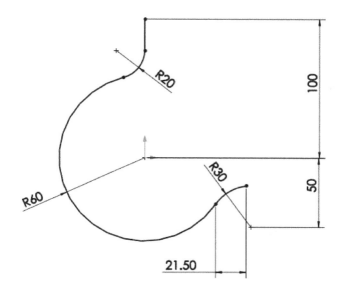

Figure-79. Dimension sketch of hook

- Exit the sketch by selecting **Exit Sketch** button from the **Ribbon**.

In previous steps, we have created the sketch of the path. Now, we will create the section sketch.

- Click on the **Plane** tool from the **Reference** drop-down in the **Features** tab of the **Ribbon**. The **Plane PropertyManager** will display as shown in Figure-80.
- Click on the top point of the line and select the **Top Plane**. Preview of the plane will display; refer to Figure-81.
- Click on the **OK** button from the **Plane PropertyManager** to create the plane.
- Select the newly created plane and select the **Sketch** button from the **Sketch** tab of the **Ribbon**. The sketching environment will display.

Figure-80. Plane PropertyManager

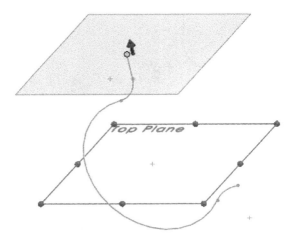

Figure-81. Preview of plane

*   Select the **Normal To** button from the **View Orientation** box. The sketching plane will become parallel to the viewport.
*   Create a circle of diameter **25** taking end point of line as center; refer to Figure-82.

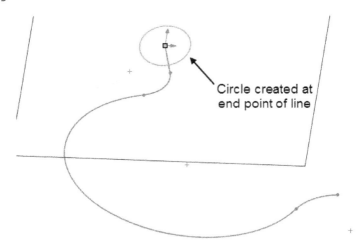

Circle created at
end point of line

Figure-82. Circle to be created

Creating the Swept Boss/Base feature

Now, we have all the sketches to create the hook. The steps to create the swept boss/base feature using these sketches is given next.

*   Click on the **Swept Boss/Base** tool from the **Ribbon**. The **Sweep PropertyManager** will display.
*   Select the circle created for section and then select sketch created for the profile.

- Click on the **OK** button from the **PropertyManager**. The swept base feature will be created; refer to Figure-83.

Figure-83. Swept feature created

Applying Conic fillet at the end

As we all know hooks doesn't end with sharp edges. So, we need to apply fillet at the end. Steps to do so are given next.

- Click on the **Fillet** tool from the **Fillet** drop-down in the **Ribbon**. The **Fillet PropertyManager** will display.
- Select the edge of the end and specify the parameters in the **PropertyManager** as shown in Figure-84.
- Click on the **OK** button from the **PropertyManager** to create the feature. The model will be displayed as shown in Figure-85.

Figure-84. Fillet at the end edge

Figure-85. Final model

Practical 3

Create the model(isometric view) as shown in Figure-86. The dimensions of the model are given in Figure-87.

Figure-86. Model for Practical3

Figure-87. Practical3 drawing views

Creating first extrude feature

- Click on the **Extrude Boss/Base** tool from the **Ribbon**. The **Boss-Extrude PropertyManager** will display.
- Select the Top Plane from the viewport and create the sketch as shown in Figure-88.

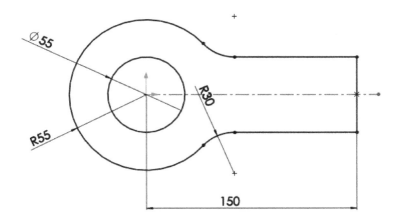

Figure-88. Sketch created at top plane

- Exit the sketch and extrude it to the height of **120**.

Creating loft feature

Before creating loft feature, we must have at least two sketch sections. First, we will create a plane at a distance of **180** from the vertical flat face of the model and then, we will create sketches and loft feature successively.

- Click on the **Plane** tool from the **Reference** drop-down and select the vertical flat face of the model.
- Specify the offset distance as **180**; refer to Figure-89.
- Click on **OK** button to create the plane.

Figure-89. Plane to be created

- Select this plane and create the sketch as shown in Figure-90.

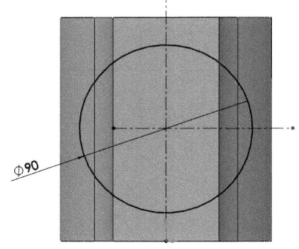

Figure-90. Sketch to created

- Exit the sketch by clicking **Exit Sketch** button.

Now, we are ready to create the Loft feature.

- Click on the **Lofted Boss/Base** tool from the **Ribbon**. The **Loft PropertyManager** will be displayed.

- Select the flat face of the model which you selected to create plane; refer to Figure-91.

Figure-91. Face to be selected

- Select the sketch created. Preview of the loft feature will be displayed; refer to Figure-92.
- Drag the green handles on the curves to manipulate the shape of lofted feature if you get it twisted.
- Click on the **OK** button from the **PropertyManager**.

Figure-92. Lofted feature to be created

Creating extrude feature

- Select the flat face of the lofted feature and click on the **Extruded Boss/Base** tool. The sketching environment will open.
- Create a circle of diameter **70** at the center and exit the sketch. Preview of extruded feature will display.
- Set the extrude depth as **220**. The preview of the model will display as shown in Figure-93.

Figure-93. Preview of model after extrusion

- Click **OK** from the **PropertyManager**.

Now, we need to create the cut feature to give front shape to model.

Creating extrude cut feature

After looking at the model, we can find that the front cut can be easily created by using the **Extruded Cut** tool.

- Click on the **Extruded Cut** tool from the **Ribbon** and select the vertical plane passing through the model i.e. Front Plane. The sketching environment will be displayed.
- Create the sketch as shown in Figure-94.

Figure-94. Sketch for cut feature

- Exit the sketch. The preview of extruded cut will display.
- Select the **Mid Plane** option from the **End Condition** drop-down and specify the height as **200**.
- Click on the **OK** button. The cut feature will be created and the model will display as shown in Figure-95.

Figure-95. Model after cut feature

Creating Fillets and Chamfers

Click on the **Fillet** tool, specify the radius as **5,** and select all the edges on which you want to create the fillet. Similarly, click on the **Chamfer** tool, specify the parameters as per the drawing and select the edges to apply the chamfer. The model after applying the fillet and chamfers is displayed as shown in Figure-96.

Figure-96. Final Model for Practical3

PRACTICE 1

Create the model as shown in Figure-97. The dimensions are given in Figure-98.

Figure-97. Practice1

SECTION A-A

Figure-98. Dimensions of the Practice1 model

PRACTICE 2

Create the model using the drawings shown in Figure-99.

Figure-99. Rope Pulley

PRACTICE 3

Create the model as shown in Figure-100. Dimensions are given in Figure-101. Assume the missing dimensions.

Figure-100. Practice3 model

Figure-101. Practice3

PRACTICE 4

Create the model by using the dimensions given in Figure-102.

Figure-102. Practice4

PRACTICE 5

Create the model by using the dimensions given in Figure-103.

Figure-103. Practice5

PRACTICE 6

Create the model by using the dimensions given in Figure-104.

Figure-104. Practice6

PRACTICE 7

Create a ring nut with value of **D** as **5,6,8,** and **10** using equation and design table. Dimensions are given in Figure-105.

Figure-105. Ring Nut

SELF ASSESSMENT

Q1. The .... tool is used to apply radius at the edges.

Q2. The .... tool is used to apply different type of fillets in one single mode.

Q3. The .... tool is used to bevel the edges of the model.

Q4. The .... tool is used to create multiple instances of a solid features along the selected path.

Q5. The .... tool is used to create multiple instances of a solid features as per the points specified in the selected sketch.

Q6. The .... tool is used to create multiple instances of a features by filling the selected bounded region.

Q7. The .... tool is used to create multiple instances of a features as per the coordinates specified in the table.

Q8. The .... tool is used to create support in the structures to increase their strength.

Q9. The .... tool is used to apply taper to the faces of a solid model.

Q10. The .... tool is used to make a solid part hollow and remove one or more faces.

For Student Notes

Chapter 6

Advanced Solid Modeling

Topics Covered

The major topics covered in this chapter are:

- *Dome tool.*
- *Freeform and Deform tools.*
- *Indent and Flex tools*
- *Split tool.*
- *Move/Copy tool.*
- *Delete/Keep Body and Convert to Mesh Body tools.*
- *Importing Bodies*
- *Fastening Features*
- *Blocks*
- *Macros*
- *Practical and Practice*

INTRODUCTION

In previous chapter, you have learned about solid editing using various tools like material removal tools, patterns, shell and so on. In this chapter, we will work with some advanced modeling tools that are used in exceptional cases. Since these tools are not generally used by an average user so they are not found in **Ribbon**. Most of the tools discussed in this chapter will be available in the Menus. We will start with advanced modeling tools in the **Features** cascading menu and then we will move forward to other menus. The tools are discussed next.

CREATING DOME

Although you can create domes in the modeling by using various surfacing tools but SolidWorks has a direct tool to reduce your efforts. The procedure to create dome is given next.

- Click on the **Dome** tool from the **Features** cascading menu of the **Insert** menu. The **Dome PropertyManager** will be displayed; refer to Figure-1.

Figure-1. Dome PropertyManager

- Select the flat faces on which you create dome feature. Preview of the features will be displayed; refer to Figure-2.
- Set the desired height for dome in the **Distance** edit box. Click on the **Reverse Direction** button if you want to remove material from the model in place of adding using the dome shape.
- If you want to use a custom direction for creating the dome then **Direction** selection box and select the desired reference.
- If you want to constrain the dome at some point then click in the **Constraint Point or Sketch** selection box and select the desired point; refer to Figure-3.

Figure-2. Preview of dome feature

Figure-3. Using point to create dome

• Click on the **OK** button from the **PropertyManager** to create the dome feature.

FREE FORMING SOLID

The **Freeform** tool in SolidWorks is used to modify the shape and size of solid model by using the control points or control polygons. The procedure to use this tool has been discussed in Chapter 9 of Surfacing.

DEFORMING SOLID

The **Deform** tool in deform a solid model using different set of instructions. The procedure to use this tool is given next.

- Click on the **Deform** tool from the **Features** cascading menu of the **Insert** menu. The **Deform PropertyManager** will be displayed; refer to Figure-4.
- Select the **Point** radio button from the **Deform Type** rollout if you want to deform the solid body using a point on it. Select the **Curve to curve** radio button if you want to deform selected edge/curve of the model to match with new one. Select the **Surface Push** radio button if you want to use triad to deforming selected face of the model.

Deforming model using Point Option

- Select the **Point** radio button from the **PropertyManager**. The options will be displayed as shown in Figure-5.

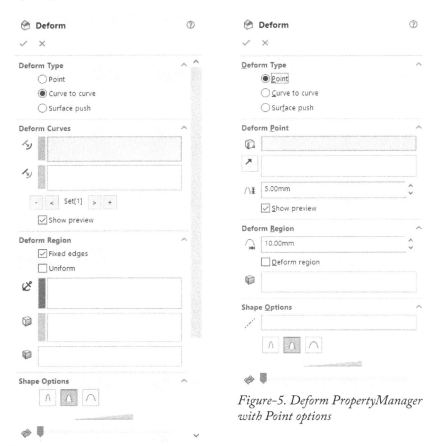

Figure-5. Deform PropertyManager with Point options

Figure-4. Deform PropertyManager

- Select a point on the model that you want to use for deformation. Preview of deformation will be displayed; refer to Figure-6.
- Set the desired value of deformation in the **Deform Distance** edit box of **Deform Point** rollout.
- Set the direction of deformation using the Deform Direction selection box as discussed earlier.
- By default, the whole body is deformed on applying this tool but if you want to confine deformation in a specific area then select the **Deform region** check box from the **Deform Region** rollout. Two selection box will be displayed below the check box.
- Click in the green selection box and select the faces that you want to be deformed.

- Using the options in Shape Options rollout, you can define the shape of deformation.
- After setting the desired values, click on the **OK** button.

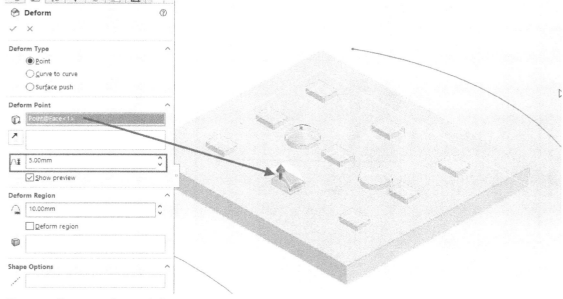

Figure-6. Preview of point deformation

Deforming Model using Curve to Curve Option

- Select the **Curve to curve** radio button from the **Deform Type** rollout. The **PropertyManager** will be displayed as shown in Figure-7.

Figure-7. Deform PropertyManager
with Curve to curve options

- Select the edge(s) that you want to be deformed.
- Click in the **Target Curves** selection box (pink colored selection box) from the **Deform Curves** rollout. You will be asked to select the target curve. Select a curve. The model will be deformed accordingly; refer to Figure-8.

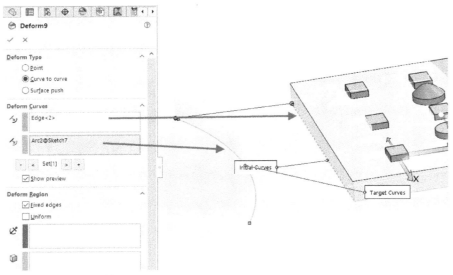

Figure-8. Deformed by curves

- To deform another set of curves, click on the **+** button in the **Deform Curves** rollout.
- After performing the desired deformation, click on the **OK** button from the **PropertyManager**.

Deforming Model using Surface push Option

- Select the **Surface push** radio button from the **Deform Type** rollout. Rest of the steps are similar to the previous two options.

CREATING INDENT ON SOLID MODEL

The **Indent** tool is used to create indent on the solid based on the selected solid face or surface. The procedure to use this tool is given next.

- Click on the **Indent** tool from the **Features** cascading menu of the **Insert** menu. The **Indent PropertyManager** will be displayed; refer to Figure-9.
- Select the body on which you want to make indent of the tool.
- Click in the **Tool body region** selection box and select the body to be used as tool for indent.
- Select the **Keep selections** radio button if you want the target body and tool body after performing indent operation.
- Select the **Remove selections** radio button if you want the target body and tool body after performing indent operation.

Figure-9. Indent Property-Manager

- By default, the indent is added to the selected body but if you want to remove the material from the target body then select the cut check box. Preview of the indent will be displayed; refer to Figure-10.

Figure-10. Preview of indent

- Click on the **OK** button to perform the operation. Note that you may need to hide the tool body manually to check the indent. To do so, select the tool body from the **FeatureManager Design Tree** and right-click on it. A shortcut menu will be displayed. Click on the **Hide** button ✎ from the shortcut menu.

APPLYING FLEX TRANSFORMATION

The **Flex** tool is used to dynamically transform a solid body using the control handles. The procedure to use this tool is given next.

- Click on the **Flex** tool from the **Features** cascading menu of the **Insert** menu. The **Flex PropertyManager** will be displayed; refer to Figure-11.

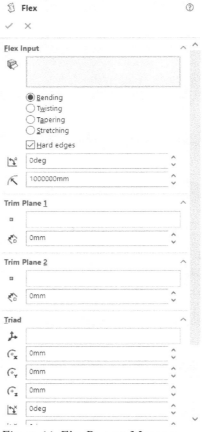

Figure-11. Flex PropertyManager

- Select the body on which you want to perform flex operations.
- Click on the desired radio button from the **Flex Input** rollout to perform bending, twisting, tapering, or stretching. The procedure to perform various operations is discussed next.

- Select the **Bending** radio button. The model will be displayed as shown in Figure-12. Note that the model can be bent only about the bend axis.
- By default, the bend axis is at the center of the model. If you want to change the location of the bend axis then you need to move and orient the triad as required. Set the desired values in the edit boxes of **Triad** rollout or you can use the handles of triad.
- Specify the desired value of angle in the **Angle** edit box of the **Flex Input** rollout. Preview of bending will be displayed; refer to Figure-13.
- Click on the **OK** button to apply bending.

Figure-12. Preview of model before bending

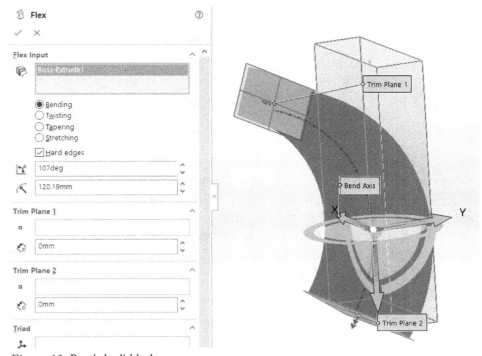

Figure-13. Bended solid body

Similarly, you can use the Twisting, Tapering and Stretching radio buttons to perform twisting, tapering and stretching of the solid bodies.

SPLITTING A SOLID BODY

The splitting of solid body is generally used in preparing mold. You will know more about molding later in this book. The procedure to perform splitting is discussed next.

- Click on the **Split** tool from the **Features** cascading menu in the **Insert** menu. The **Split PropertyManager** will be displayed; refer to Figure-14.

Figure-14. Split PropertyManager

- Select the plane or surface by which you want to split the solid and click on the **Cut Part** button from the **PropertyManager**. Based on your selection, the split-up bodies will be displayed in the preview and **PropertyManager**; refer to Figure-15.
- Click on the **OK** button from the **PropertyManager** to create split-up bodies. Note that the bodies can be found in the **Solid Bodies** folder of the **FeatureManager Design Tree**.

Figure-15. Preview of split-up bodies

MOVING AND COPYING OBJECTS

The **Move/Copy** tool is used to create copies of the selected body or move the bodies using constraint. Using this tool, you can also move the bodies dynamically using a triad. The procedure to use this tool is given next.

- Click on the **Move/Copy** tool from the **Features** cascading menu of the **Insert** menu. The **Move/Copy Body PropertyManager** will be displayed; refer to Figure-16.
- Select the body/bodies that you want to move or copy.
- Click on the desired constraint button from the **Mate Settings** rollout. Now, click in the selection box of Mate Settings rollout and select the faces on which you want to apply the constraints.

Or

- Click on the **Translate/Rotate** button at the bottom in the **PropertyManager**. The **PropertyManager** will be displayed as shown in Figure-17.

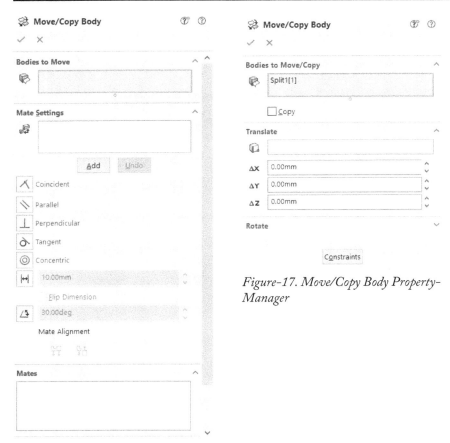

Figure-16. Move/Copy Body Property-
Manager

Figure-17. Move/Copy Body Property-
Manager

- Select the **Copy** check box if you want to create a copy of the body while moving it.
- Enter the desired values in the edit boxes of the **Translate** rollout or drag the center point of the body at desired location; refer to Figure-18.

Figure-18. Moving solid body

- Click on the **OK** button from the **PropertyManager** to apply the changes.

DELETING BODIES

The **Delete/Keep Body** tool is used to delete selected bodies. The procedure to do so is given next.

- Click on the **Delete/Keep Body** tool from the **Features** cascading menu of the **Insert** menu. The **Delete/Keep Body PropertyManager** will be displayed; refer to Figure-19.

Figure-19. Delete/Keep Body PropertyManager

- Select the **Delete Bodies** radio button if you want the selected bodies to deleted and rest of the bodies will be kept. Select the **Keep Bodies** radio button if you want the selected body to be kept and rest of the bodies to be deleted.
- Select the desired bodies to be deleted or kept.
- Click on the **OK** button to apply the changes.

CONVERTING TO MESH BODIES

A mesh is construction of polygons. The procedure to convert model into mesh is given next.

- Click on the **Convert to Mesh Body** tool from the **Features** cascading menu of the **Insert** menu. The **Convert to Mesh Body PropertyManager** will be displayed; refer to Figure-20.

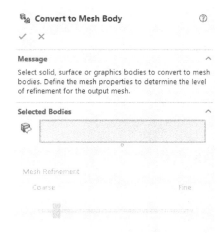

Figure-20. Convert to Mesh Body PropertyManager

- Select the body that you want to convert into mesh. The body will be displayed as mesh; refer to Figure-21.

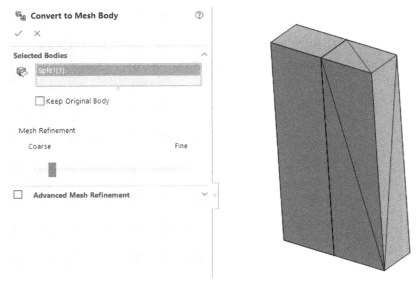

Figure-21. Preview of mesh

- Move the slider towards Coarse or Fine to refine the mesh.
- If you want to define the element size and other parameters manually then select the **Advanced Mesh Refinement** check box and then select the **Define Maximum Element Size** check box. The options will be displayed as shown in Figure-22.

Figure-22. Options for advanced refinement

- Set the desired element size and click on the **OK** button to create mesh.

IMPORTING IN PART ENVIRONMENT

The procedure to import model in Part environment is given next.

- Click on the **Imported** tool from the **Features** cascading menu in the **Insert** menu. The **Open** dialog box will be displayed.
- Select the desired model file and click on the **Open** button. The part model will be imported at the default origin; refer to Figure-23.

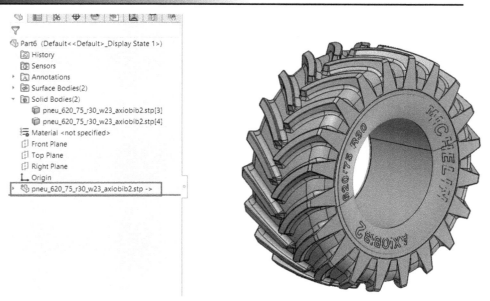

Figure-23. Imported part

FASTENING FEATURES

The Fastening features of SolidWorks are useful when creating plastic molded parts for assembly. These features include mounting boss, snap hook, snap hook groove, and so on. The procedures to create these features are given next.

Creating Mounting Boss

The Mounting boss is used to assemble two components while providing location stability. The procedure to create mounting boss is given next.

- Click on the **Mounting Boss** tool from the **Fastening Feature** cascading menu of the **Insert** menu. The **Mounting Boss PropertyManager** will be displayed; refer to Figure-24.
- Select the face which you want to create the mounting boss. Preview of the mounting boss will be displayed.
- Click in the **Select Direction** selection box and select the reference for defining the direction of mounting boss.
- If you want to place the mounting boss on a circular edge then click in the **Circular Edge** selection box in the **Position** rollout and select the round edge on which you want the mounting boss to be created.

Figure-24. Mounting Boss Property-Manager

- Select the desired type of boss from the **Boss Type** rollout and specify the desired dimensions of the boss.
- After setting the desired parameters, click on the **OK** button to create mounting boss.

Creating Snap Hook

The **Snap Hook** tool is used to create snap hook for plastic parts. The procedure to do so is given next.

- Click on the **Snap Hook** tool from the **Fastening Features** cascading menu of the **Insert** menu. The **Snap Hook PropertyManager** will be displayed; refer to Figure-25.
- Select the edge on which you want to create snap hook. Preview of the snap hook will be displayed; refer to Figure-26.

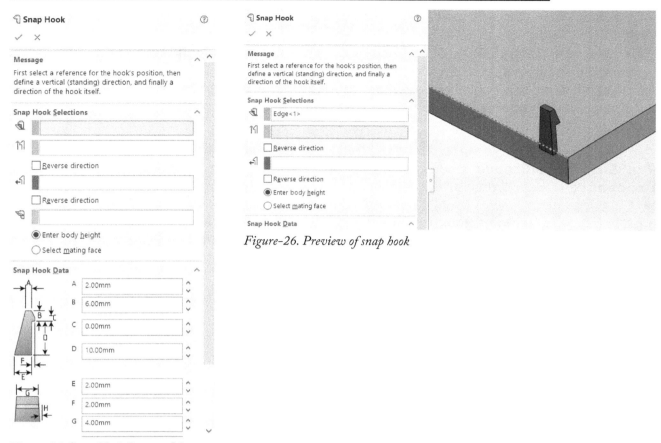

Figure-25. Snap Hook PropertyManager

Figure-26. Preview of snap hook

- Click in the pink selection box and select the desired reference for vertical direction. Select the **Reverse direction** check box to flip the direction.
- Click in the violet colored selection box and select the reference to define horizontal direction of the snap hook. The snap hook will be displayed as in Figure-27.

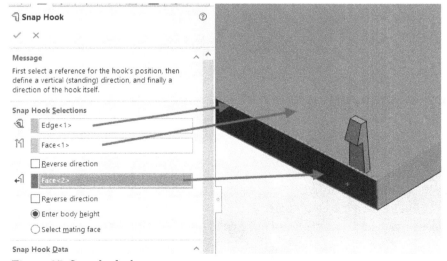

Figure-27. Snap hook placement

- Select the **Select mating face** radio button and select the desired face if you want the height of snap hook to be decided based on the selected face. Otherwise select the **Enter body height** radio button and specify the snap hook height in the respective edit box.
- Set the other parameters as required and click on the **OK** button from the **PropertyManager** to create the snap hook.

Creating Snap Hook Groove

The **Snap Hook Groove** tool is used to create groove for snap hook. The procedure to create snap hook groove is given next.

- Click on the **Snap Hook Groove** tool from the **Fastening Feature** cascading menu of the **Insert** menu. The **Snap Hook Groove PropertyManager** will be displayed; refer to Figure-28.

Figure-28. Snap Hook Groove Property-tyManager

- Select the snap hook for which you want to create the groove. You will be asked to select the body on which the groove is to be created.
- Select the body on which you want to create groove. Note that this body should be different from the one on which the snap hook is created and it must intersect with the snap hook; refer to Figure-29. On selecting the body, preview of the snap hook groove will be displayed; refer to Figure-30.

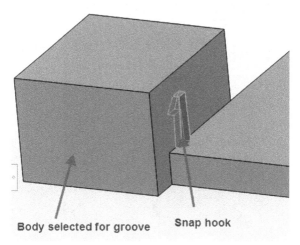

Body selected for groove   Snap hook

Figure-29. Selection for snap hook groove

Figure-30. Preview of snap hook groove

- Set the parameters as required in the **PropertyManager** and click on the **OK** button to create the groove.

Creating Vent

Vent is used to provide cut for air flow. The procedure to create vent is given next.

- Click on the **Vent** tool from the **Fastening Feature** cascading menu of the **Insert** menu. The **Vent PropertyManager** will be displayed; refer to Figure-31.

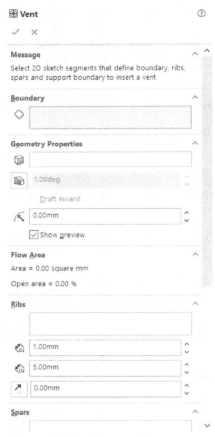

Figure-31. Vent PropertyManager

The procedure to create vent has been discussed later in this book.

Creating Lip and Groove

Lip and Groove is used to guide assembly of plastic parts. The procedure to create lip or groove is given next.

* Click on the **Lip/Groove** tool from the **Fastening Feature** cascading menu of the **Insert** menu. The **Lip/Groove PropertyManager** will be displayed; refer to Figure-32.

Figure-32. Lip Groove PropertyManager

* Select the body on which you want to create groove. You will be asked to select the body on which the lip is to be created.

- Select the other body. You will be asked to select the reference plane/face/edge/axis for direction of lip and groove mate.
- Select the desired reference for direction; refer to Figure-33. The body selected for lip will hide automatically and you will be asked to select the face of other body to create groove.

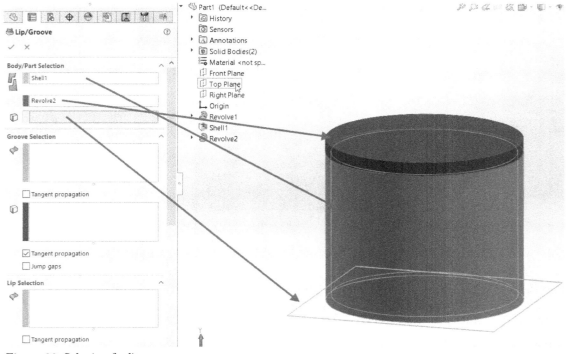

Figure-33. Selection for lip groove

- Select the face where this body meets the other.
- Click in the next selection box of the **Groove Selection** rollout. You will be asked to select the edge from where the groove should start.
- Select the desired edge. Preview of the groove will be displayed; refer to Figure-34.

Figure-34. Preview of the groove

- Note that there is a gap in the groove due to boss feature in the body. If you want to create groove through this gap then select the **Jump Gaps** check box. Preview of the groove will be displayed accordingly; refer to Figure-35.

Figure-35. Preview of the groove with Jump gaps check box selected

- Click in the first selection box of the **Lip Selection** rollout. The body selected for lip will be displayed and the other body will hide automatically. Select the face mating to the other body.
- Click in the next edit box of the **Lip Selection** rollout. You will be asked to select the edge on which the lip feature is to be created.
- Select the desired edge. Preview of the lip feature will be displayed; refer to Figure-36.

Figure-36. Preview of lip feature

- Set the desired parameters in the **PropertyManager** and click on the **OK** button to create the feature.

BLOCK DESIGNING

The block designing is a powerful method to group the objects and perform various manipulations. The tools to create block in **Part** environment are available in the **Blocks** cascading menu of the **Tools** menu; refer to Figure-37.

Figure-37. Blocks cascading menu

Note that the tools in this menu are active when you are editing or creating a block.

Creating Block Using Sketch

The blocks are created in sketching environment. The procedure to do so is given next.

- Create a sketch of the model that you want to use as block; refer to Figure-38.

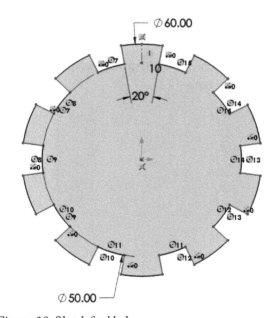

Figure-38. Sketch for block

- Make sure you are still in the sketching environment and then select the **Make** tool from the **Blocks** cascading menu of the **Tools** menu. The **Make Block PropertyManager** will be displayed; refer to Figure-39.

Figure-39. Make Block PropertyManager

- Select the entities of the sketch to be included in the block.
- To define the insertion point for the block, expand the **Insertion Point** rollout and drag the blue colored coordinate system at desired location; refer to Figure-40.

Figure-40. Dragging the insertion point

- Click on the **OK** button from the **PropertyManager** to convert the sketch into block; refer to Figure-41.

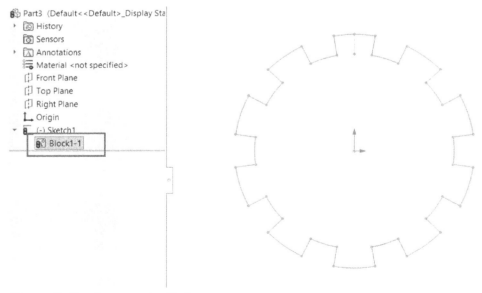

Figure-41. Sketch converted to block

Saving the Block

After making the block as discussed earlier, click on the **Save** tool from the **Blocks** cascading menu of the **Tools** menu. (Make sure you do not leave the sketching environment). The **Save As** dialog box will be displayed. Specify the desired name and location for the file and click on the **Save** button to save the block.

Editing a Block

Editing a block involves sketching tools and block toolbar. The procedure to edit a block is given next.

- To edit a block, select it from the **FeatureManager Design Tree** and right-click on it. A shortcut menu will be displayed; refer to Figure-42. Click on the **Edit Block** option from the shortcut menu. The **Sketching** environment will be displayed with **Blocks** tool bar.

Figure-42. Edit block option

Figure-43. Editing a block

- Edit the block as required by using the sketching tools.
- If you want to add/remove anything from the block then click on the **Add/Remove** button from the **Blocks** toolbar. The **Add/Remove Entities PropertyManager** will be displayed; refer to Figure-44.

Figure-44. Add Remove Entities PropertyManager

- Select the entities that you want to be added in the block from the drawing area.
- To remove an entity from the block, select it from the list. Right-click on it and select the **Delete** option.
- Click on the **OK** button from the **PropertyManager** to apply the changes.
- Now, click on the **Rebuild** button tool from the **Blocks** toolbar to update the block.
- Click on the **Save Block** tool from the toolbar to save the block.

Exploding Block

The **Explode Block** tool is used to break the block into sketch entities. To do so, select the block from the **FeatureManager Design Tree** and right-click on it. A shortcut menu will be displayed. Click on the **Explode Block** tool from the shortcut menu. The block will be converted to a sketch.

MACROS

Macro is a set of instructions to be run by computer automatically once giving a call. Macros can reduce the modeling time at a greater extent if used efficiently. The tools to create and manage macros are available in the **Macro** cascading menu of the **Tools** menu. These tools are discussed next.

Figure-45. Macro cascading menu

Creating Macros

The procedure to create macro is given next.

- Click on the **New** tool from the **Macro** cascading menu of the **Tools** menu. The **Save As** dialog box will be displayed; refer to Figure-46.

Figure-46. Save As dialog box

- Specify the desired name and save the file at desired location by clicking the **Save** button. The Microsoft Visual Basic interface will be displayed with the macro file open for editing; refer to Figure-47.

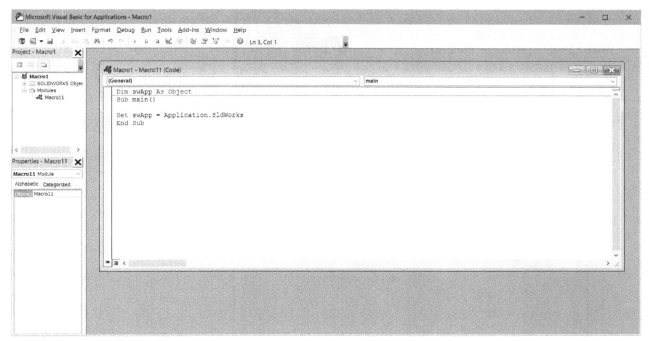

Figure-47. Microsoft Visual Basic interface

- Create the desired program using visual basic codes and API of SolidWorks.
- After creating the program, save the file and close the application window.

Note that you need to have working knowledge of Microsoft VBA to create codes for SolidWorks macros. Apart from this hard way, there is an easier way to create macro by recording it while working in SolidWorks.

Recording Macros in SolidWorks

The **Record** tool in **Macro** cascading menu is used to record macros. The procedure is discussed next.

• Click on the **Record** tool from the **Macro** cascading menu of the **Tools** menu. The **Macro** toolbar will be added above the **Ribbon**; refer to Figure-48.

Figure-48. Macro toolbar

• Perform the steps as required and then click on the **Stop** button. The **Save As** dialog box will be displayed.
• Specify the desired name and click on the **Save** button. The macro will be created.

Running a Macro

The **Run** tool in the **Macro** cascading menu is used to run recorded macros. The procedure is given next.

• Click on the **Run** tool from the **Macro** cascading menu of the **Tools** menu. The **Open** dialog box will be displayed; refer to Figure-49.

Figure-49. Open dialog box

• Double-click on the macro that you want to run. The macro will run automatically.

Editing Macro

The **Edit** tool in the **Macro** cascading menu is used to edit earlier created macro. The procedure to edit the macro is given next.

- Click on the **Edit** tool from the **Macro** cascading menu of the **Tools** menu. The **Open** dialog box will be displayed.
- Double-click on the macro file that you want to edit. The Microsoft VBA application will be displayed with the file opened; refer to Figure-50.

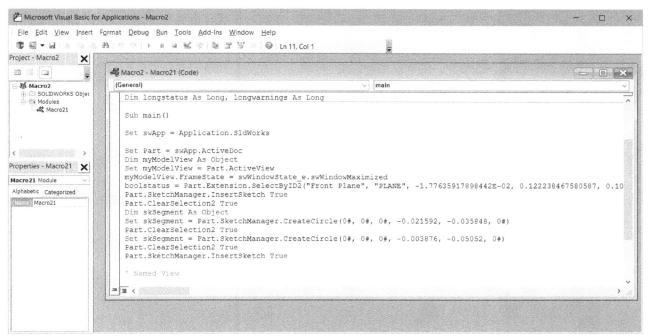

Figure-50. Macro file opened

- Edit the macro and save the file.

FOR STUDENT NOTES

Chapter 7

Assembly and Motion Study

Topics Covered

The major topics covered in this chapter are:

- *Inserting Components in Assembly.*
- *Assembly Constraints.*
- *Reference and Assembly features.*
- *Exploded View.*
- *Bill of Material.*
- *Motion Study.*

Edit Component   Insert Components   Mate   Component Preview Window   Linear Component Pattern   Smart Fasteners   Move Component   Show Hidden Components   Assembl...   Referenc...   New Motion Study   Bill of Materials   Expl Vi

Assembly | Layout | Sketch | Evaluate | Render Tools | SOLIDWORKS Add-Ins | SOLIDWORKS MBD

ASSEMBLY

In engineer's language, assembly is the combination of two or more components and these components are constrained to each other in a specified manner called assembly constraints. In SolidWorks, Assembly Design is a separate environment. To start the Assembly Design, click on the **New** button from the **Menu Bar**. The **New SolidWorks Document** dialog box will display; refer to Figure-1. Double-click on the **Assembly** button from the dialog box. The Assembly Design environment will display as shown in Figure-2. The tools related to assembly are available in the **Ribbon**. Note that in the left of the screen, the **Begin Assembly PropertyManager** is displayed. In SolidWorks 2017, the **Open** dialog box is also displayed prompting you to select file(s) to be inserted in the assembly.

Figure-1. New SolidWorks Document dialog box

Figure-2. Assembly Design environment

INSERTING BASE COMPONENT

- After starting Assembly environment, select the desired part file from the **Open** dialog box and click on the **Open** button. The part will get attached to the cursor; refer to Figure-3.

Figure-3. Part attached to cursor

- Specify the rotation angle value in the edit box and click on the desired direction button from the dynamic input box to rotate the part.
- Click on the **OK** button from the **PropertyManager** to place the part. Note that the base component of assembly is inserted as fixed in SolidWorks and **(f)** is displayed next to its name in the **FeatureManager Design Tree**; refer to Figure-4.

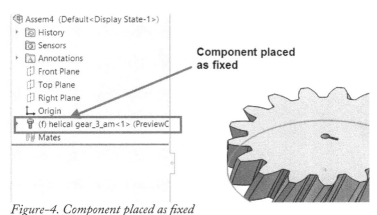

Figure-4. Component placed as fixed

INSERTING COMPONENTS IN ASSEMBLY

Once, you have inserted the base component in the assembly, follow the steps given next to insert more components.

- Click on the **Insert Components** tool from the **Assembly CommandManager** in the **Ribbon**. The **Open** dialog box will display; refer to Figure-5.
- Browse to the location of your desired part file and double-click on the file to add it. The component will be attached to the cursor; refer to Figure-6. You can rotate the component by using the **Rotate** options in dynamic input box.

- Click in the viewport to place the component. The component will be inserted in the assembly.

Figure-5. Open dialog box

Figure-6. Component attached to cursor

- The component will be placed as floating and can move anywhere by dragging. Also **(-)** mark will be added before its name in the **FeatureManager Design Tree** displayed at the left.
- You can use the **Rotate** menu to rotate the part before inserting.

- Similarly, you can insert more components as per your requirement in the assembly.
- If you want to reinsert a component then hold the CTRL key from keyboard and drag the desired component already existing in the display area. One more instance of component will be inserted; refer to Figure-7.

Figure-7. Reinserting a component

- From SolidWorks 2016 onwards, you can also choose the desired configuration of the part being inserted in assembly. To do so, click on the **Configuration** drop-down in the **Insert Component PropertyManager** and select the desired configuration; refer to Figure-8.

Figure-8. Configuration drop-down

After adding all the components in assembly, the next task is to apply proper constraints to them. The next section explains the use of assembly constraints (also called **Mates** in the language of SolidWorks).

ASSEMBLY CONSTRAINTS (MATES)

The options to apply assembly constraints are available in the **Mate PropertyManager**. The use of options in this **PropertyManager** is explained next.

- After inserting the desired components, click on the **Mate** tool from the **Ribbon**. The **Mate PropertyManager** will display as shown in Figure-9.
- From SolidWorks 2016 onwards, you can make the first selection for mate as transparent for easy identification. To do so, scroll down in **Mate PropertyManager** and select the **Make first selection transparent** check box; refer to Figure-10.

Figure-9. Mate PropertyManager

Figure-10. First selection transparency option

- If you want to position the components by using the mates but do not want to apply the mates then select the **Use for positioning only** check box from the **Options** rollout in the **PropertyManager**.
- In SolidWorks 2018, you can now hide or show faces while applying mates. To hide a face of model, move the cursor over desired face and press **ALT** from the keyboard. To display a hidden face again, press **SHIFT**+**ALT** from the keyboard when the cursor is over hidden surface area.

The buttons used for applying constraints are discussed next.

Coincident

The **Coincident** button is used to make two components coincide at the selected references. The steps to use this constraint are given next.

- Click on the **Coincident** button from the **PropertyManager.** You are asked to select two references for this mate.
- Select two axes/faces/planes/curves that you want to make coincident. The preview of constraint will display with a pop-up toolbar; refer to Figure-11.

Figure-11. Coincident constraint preview

- Click on the **Flip** button from pop-up toolbar to change the orientation of components; refer to Figure-12.

Figure-12. Components after flipping

Parallel

The **Parallel** button is used to make two components parallel with respect to each other. The steps to use this button are given next.

- Click on the **Parallel** button and select the two faces/axes/planes. The components will be parallel with respect to these references. Refer to Figure-13.

Figure-13. Parallel constraints

Note that the tools displayed in the pop-up toolbar are also in the **PropertyManager** and work in the same way. The pop-up toolbar gives us facility to change the constraint type after the command is activated.

Perpendicular

The **Perpendicular** button is used to make two components perpendicular to each other. The steps are given next.

- Click on the **Perpendicular** button and select two faces/axes/plane that you want to make perpendicular to each other. The preview will display; refer to Figure-14.

Figure-14. Perpendicular constraint

Tangent

The **Tangent** button is used to make two components tangent to each other. The steps are given next

- Click on the **Tangent** button and select two faces that you want to make tangent. The preview of tangent constraint will display as shown in Figure-15.

Figure-15. Tangent constraint

Concentric

The **Concentric** button is used to make two round components share the same center axis. The steps are given next.

- Click on the **Concentric** button and select two round faces that you want to make concentric. The preview of tangent constraint will display as shown Figure-16.

Figure-16. Concentric constraint

Note that you can lock the rotation of selected components by selecting the **Lock Rotation** check box in the **PropertyManager**/pop-up toolbar. In SolidWorks 2018, if there is a small misalignment between two holes while making them concentric then the mate toolbar is displayed as shown in Figure-17.

Figure-17. Mate toolbar for mis-aligned concentric

Click on the **Misaligned** button. The constraint will be applied and the deviation will be displayed in the **Result** box of **Mate PropertyManager**; refer to Figure-18.

Lock

The **Lock** button is used to lock the component at its current position. The steps to do so are given next.

- Click on the **Lock** button and select the components you want to fix. The preview of components will display in blue color.
- Click **OK** to fix the component. Now, drag one of component to check the effect.

Note that you cannot move the first component of assembly by default. To move the first component, right click on its name in the **FeatureManager Design Tree** and select the **Float** option from the menu displayed; refer to Figure-19. Now, you will be able to move the component.

Figure-18. Concentric constraint with deviation

Figure-19. Float option in shortcut menu

Distance

The **Distance** button is used to set distance between two selected faces. The steps are given next.

- Click on the **Distance** button and select two flat faces that you want to use for setting distance. The preview of distance constraint will display; refer to Figure-20.

Figure-20. Distance constraints

- Set the desired distance in the pop-up toolbar and select the **Flip** button if required.

Angle

The **Angle** button is used to set angle between two selected faces. The steps are given next.

- Click on the **Angle** button and select two flat faces that you want to use for setting angle. The preview of angle constraint will display; refer to Figure-21.

Figure-21. Angle constraint

- Set the desired angle in the pop-up toolbar and select the **Flip** button if required.

Note that if you have applied any wrong constraint and want to delete it then click on **+** sign next to **Mates** in the **FeatureManager Design Tree**, select the desired constraint from the list and press **DEL** button from keyboard.

Till this point, we have learned **Standard Mates** that are used for rigid assemblies. Now, we will move to advanced constraints (Mates) that play key role in motion study. Note that you might need to delete all the standard mates to apply advanced and mechanical mates for checking motion.

Expand the **Advanced Mates** rollout from the **FeatureManager** to display the advanced mates; refer to Figure-22. The mates in this rollout are explained next.

Figure-22. Advanced Mates

Profile Center

The **Profile Center** button is used to align the two components at a common center of the faces selected for mate. The steps to use this mate are given next.

- Click on the **Profile Center** button and select the flat faces of the components. The faces will be align at their profile centers; refer to Figure-23.
- Specify the desired distance between the faces by using the edit box displayed below the **Profile Center** button in the **PropertyManager**.

Figure-23. Applying Profile center mate

Symmetric

The **Symmetric** button is used to make faces of components symmetric with respect to a reference. In other words, the distance by which one component move will be the same as distance moved by other symmetric component. The steps are given next.

- Click on the **Symmetric** button and select two flat faces that you want to make symmetric.
- Click in the **Symmetry plane** selection box in the **Mate Selections** rollout and select the plane about which you want to make the components symmetric. The preview of applied mate will display; refer to Figure-24.

Figure-24. Symmetric constraint

Width

The **Width** button is used to fit a component in the selected width reference. The steps to use this mate are given next.

- Click on the **Width** button and select two flat faces in the assembly that define the width reference.
- Now, select the two flat faces of your components to defined total width of your component. The preview of mate will display; refer to Figure-25.

Figure-25. Width mate

Note that if the width of references is more than the limits of component then the component will be inserted in the middle of the width references.

Path Mate

The **Path Mate** button is used to make pointed component follow the specified path. The steps to use this mate are given next.

- Click on the **Path Mate** button and select the vertex of the component that you want to make follower.
- Select sketch of the path that you want to make as guide for the follower. The preview of path mate will display; refer to Figure-26.

Figure-26. Path mate constraint

- You can change the options related to movement of object by using the drop-down and edit boxes displayed below **Path Mate** in **PropertyManager**; refer to Figure-27.

Figure-27. Options for path mate

- If you want to move the component at certain distance or path length percentage then select the respective option from the **Path Constraint** drop-down in the **PropertyManager** and specify the related value; refer to Figure-28.

Figure-28. Using Path Constraint drop-down

- By default, the vertex in path mate follows the path but the part is free to rotate about the vertex. To fix the orientation of part, select the **Up Vector** option from the **Roll Control** drop-down in the **PropertyManager**. The options for up vector will be displayed as shown in Figure-29.

Figure-29. Up Vector options

- Select an edge/plane/axis from the assembly to reference up vector. Select the desired direction radio button from the **PropertyManager**. The follower will be aligned accordingly; refer to Figure-30.

After setting up-vector   **Before setting up-vector**

Figure-30. Setting Up Vector for follower part

Linear/Linear Coupler

The **Linear/Linear Coupler** button is used to make two components move with respect to each other by a specified ratio. The steps to use this mate are given next.

- Click on the **Linear/Linear Coupler** button and select the faces of the component between which you want to apply the mate. Preview of mate will display; refer to Figure-31.
- Set the desired ratio between the components. Click **OK** to apply the mate.
- To check the motion, drag one of the component in linear direction. The other component will move automatically.

Figure-31. Linear coupler mate

Advanced Distance

The **Advanced Distance** button is used to apply maximum and minimum movement limit of a component. The steps to apply this mate are given next.

- Click on the **Distance** button from the **Advanced Mates** rollout. The options in the rollout will display as shown in Figure-32.
- Select the face/plane of component and then select the face/plane of reference to limit the movement. Preview of mate will display as shown in Figure-33.

Figure-32. Advanced distance mate options

Figure-33. Planes selected for advanced distance

Advanced Angle

The **Advanced Angle** button is used to apply maximum and minimum rotation limit of a component. This mate works in the way similar to **Advanced Distance** mate.

Till this point, we have learned Standard Mates and Advanced Mates. There are also a few mates that represent mechanical motion in assembly. These constraints are grouped in a rollout named **Mechanical Mates**. To use these mates, expand the **Mechanical Mates** rollout; refer to Figure-34. The buttons in this rollout are explained next.

Figure-34. Mechanical Mates

Cam

The **Cam** button is used to create cam-follower mate between two entities. The steps given next explain the procedure.

- Click on the **Cam** button from the **Mechanical Mates** rollout. The selection boxes will be displayed as shown in Figure-35.

Figure-35. Cam Mate

- Follow the steps in the above figure. Figure-36 shows an example of cam mate.

Note that you need to limit the motion of follower so that it move only in vertical direction.

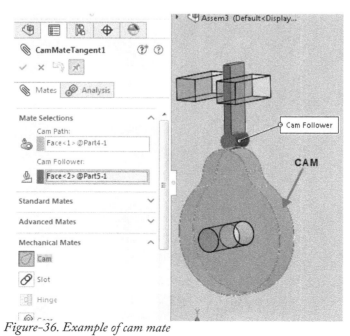

Figure-36. Example of cam mate

Slot

The **Slot** button is used to create slot-follower mate between two entities. After applying this mate, the component will move only in the limits of slot. The steps given next explain the procedure.

• Click on the **Slot** button from the **Mechanical Mates** rollout.
• Select the round face of slot and the follower, the mate will be applied.
• You can set the starting position of follower by using the options in the **Constraint** drop-down displayed below the **Slot** mate in **PropertyManager**. Figure-37 shows an example of slot mate.

Figure-37. Example of slot mate

Hinge

The **Hinge** button is used to make the two parts behave as hinged. The procedure to use this button is given next.

• Click on the **Hinge** button from the **Mechanical Mates** rollout. The selection boxes in the **Mate Selections** rollout are displayed as shown in Figure-38.

Figure-38. Hinge Mate Selection boxes

- Select the two round faces that you want to be concentric.
- Select the two flat faces that define the angular motion limit of the hinge.
- Select the **Specify angle limits** check box and specify the limits as explained in advanced mates. Figure-39 shows a hinge mate being applied to the assembly.

Coincident faces

Concentric faces

Figure-39. Example of hinge mate

Gear

The **Gear** button is used to create a joint between two gears. The procedure to use this button is given next.

- Click on the **Gear** button from the **Mechanical Mates** rollout.
- Select the flat faces of the gears. The **Gear Mate PropertyManager** will display as shown in Figure-40. Also, the gears will display as shown in Figure-41.

Figure-40. Gear Mate Property-tyManager

Figure-41. Example of gear mate

Rack Pinion

The **Rack Pinion** button is used to create joint between a rack and a pinion. The procedure to use this button is given next.

- Click on the **Rack Pinion** button from the **Mechanical Mates** rollout.
- Select edge of the rack and flat face of the pinion gear. The rack and pinion mate will be created; refer to Figure-42. Specify the desired ratio in the **PropertyManager**.

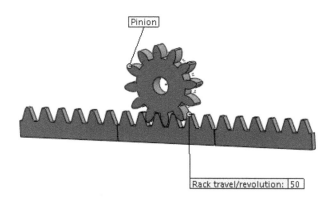

Figure-42. Rack pinion mate

Screw

The **Screw** button is used to create screw joint between a cylinder and a flat face. The procedure to use this button is given next.

- Click on the **Screw** button from the **Mechanical Mates** rollout.
- Select the cylindrical face of first component and flat face of other component. The preview will be displayed; refer to Figure-43.
- Specify the desired value in the box displayed below **Screw** button in the **PropertyManager**.

Figure-43. Screw Mate

Universal Joint

The **Universal Joint** button is used to create universal joint between two components. After clicking this button, select the two round faces. The universal joint will be created; refer to Figure-44.

Figure-44. Universal joint

EXPLODED VIEW

After creating assembly, we are required to display the components of assembly expanded in the way they are assembled. Follow the steps given below to create the exploded view of assembly.

* Select all the components that you want to explode from the main assembly and select the **Exploded View** tool from the **Ribbon**. The **Explode PropertyManager** will display; refer to Figure-45.

Figure-45. Explode PropertyManager

* Select the **Auto-space components on drag** check box from the **Options** rollout in the **Explode PropertyManager** and drag the components in the desired direction using the arrow handles.
* Click on the **OK** button from the **PropertyManager**. The components will be exploded automatically.

Or

* Click on the **Exploded View** button and select the component you want to move. A triad will display on the component.
* Drag the component using the handles displayed to the desired distance or specify the desired distance and rotation in the edit boxes available in the **Settings** rollout of the **PropertyManager**.
* Repeat the above steps until you explode all the components.

To display the explode lines, click on the **Explode Line Sketch** button. The **Route Line PropertyManager** will display; refer to Figure-46.

Figure-46. Route Line PropertyManager

- Click on the assembly reference of the component and then the corresponding assembly reference of the base component. The explode line will be created; refer to Figure-47.

Figure-47. Explode lines created

Smart Explode Lines

The **Smart Explode Lines** tool is used to create explode lines automatically while taking center of bounding box as reference. The procedure to use this tool is given next.

- Select the exploded view from **ConfigurationManager** and right-click on it. A shortcut menu will be displayed; refer to Figure-48.
- Click on the **Smart Explode Lines** tool from the shortcut menu. The **Smart Explode Lines PropertyManager** will be displayed; refer to Figure-49.

Figure-48. Shortcut menu for exploded view

Figure-49. Smart Explode Lines PropertyManager

- Select the desired option from the **Component route line** rollout. If you want to use the bounding box center as reference for explode lines. Similarly, you can use the other options.
- After setting the desired parameters, click on the **OK** button from the **PropertyManager** to create the lines.

BILL OF MATERIALS

Bill of materials is used to list the total components of the assembly in the form of a table. To create bill of materials, follow the steps given below.

- Click on the **Bill of Materials** button from the **Ribbon**. The **Bill of Materials PropertyManager** will display; refer to Figure-50.

Figure-50. Bill of Materials PropertyManager

- Click on the [image] button to display templates. The **Open** dialog box will be displayed with default templates of bill of materials; refer to Figure-51.

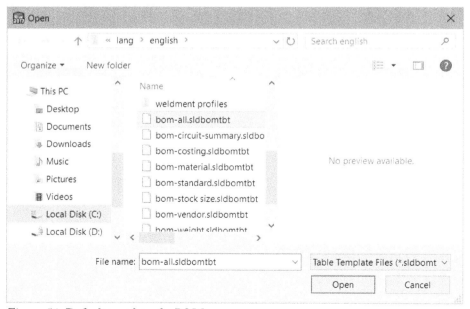

Figure-51. Default templates for BOM

- Select the desired template from the dialog box and click on the **Open** button. The template will be activated.
- Click on the **OK** button from the **PropertyManager**. You are asked to place the table of bill of materials.

- Click **OK** from the **Select Annotation View** dialog box.
- Click in the viewport to place the table; refer to Figure-52.

ITEM NO.	PART NUMBER	DESCRIPTION	WEIGHT	QTY.
1	Body			1
2	Washer			1
3	Valve spindle			1
4	Spring			1
5	Spring cap nut			1
6	Pre adjusting nut			1
7	Nozzle valve			1
8	Nozzle body			1
9	Cap nut			2
10	atomiser			1

Figure-52. Bill of materials

MATE CONTROLLER

The **Mate Controller** tool was added in SolidWorks 2016. Using this tool, you can manage the distance and angle mates applied in the assembly through a single option box. The procedure to use this tool is given next.

- Click on the **Mate Controller** tool from the **Insert** menu; refer to Figure-53. The **Mate Controller PropertyManager** will be displayed; refer to Figure-54.

Figure-53. Mate Controller tool

Figure-54. Mate Controller Proper-tyManager

- Click on the **Collect All Supported Mates** button from the **Mates** rollout of the **PropertyManager**. The mates that can be controlled with their angular or distance dimension will be displayed in the selection box; refer to Figure-55.

Figure-55. Mates selected automatically

- Set the desired dimension spinners available in the **Mates Positions** rollout.
- You can save the positions by using the **Add Position** 🖳 button from the **Mates Positions** rollout. On doing so, the **Name Position** dialog box will be displayed; refer to Figure-56.

Figure-56. New position of mates

- Specify the desired name and click on the **OK** button from the dialog box.
- Click on the **Add Configuration** button 🖳 to make a new configuration of assembly at current position.
- You can select the desired position from the drop-down in the **Mates Positions** rollout; refer to Figure-57.

Figure-57. Positions rollout

- To animate the position change, expand the **Animation** rollout in the **PropertyManager** and click on the **Calculation Animation** button from it; refer to Figure-58.

Figure-58. Animation rollout

- After calculation, you can use the **Play** button from the **Animation** rollout to play the animation.
- Click on the **OK** button from the **PropertyManager** to apply the mate control. The Mate Controller feature will be added in the **FeatureManager Design Tree**.

MOTION STUDY

Motion study is used to check the motion of components with respect to each other after applying the driving force the components. This feature of SolidWorks help to understand the mechanism of our assembly in real world conditions. To start the motion study, click on the **Motion Study 'x'** tab displayed at the bottom bar of the viewport. Note that here **'x'** is the sequence number. After clicking on this tab, the interface is displayed as shown in Figure-59.

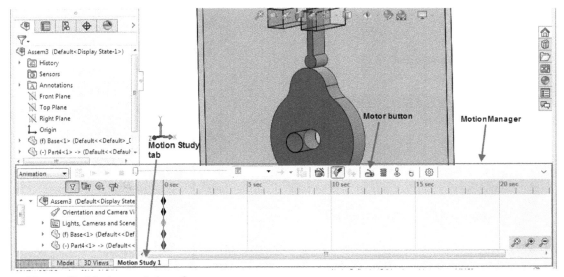

Figure-59. Motion study interface

Now, we need to apply some driving force to one component so that we can check the motion of other parts. In this case, we are using the cam-follower mechanism as displayed above. We need to add the rotary motion to the cam to check motion of follower. The steps to do so are given next.

• Click on the **Motor** button shown in the above figure. The **Motor PropertyManager** will display; refer to Figure-60.

Figure-60. Motor PropertyManager

• Click on the circular face of the part to apply the rotary motion; refer to Figure-61.

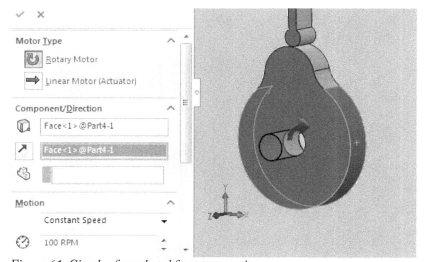

Figure-61. Circular face selected for rotary motion

• Specify the desired options for speed in the **Motion** rollout and click **OK** to apply the motion. The motor will be added in the **MotionManager**.

Note that in the same way you can apply linear motion to a flat face by selecting **Linear Motor (Actuator)** button from the **Motor PropertyManager** in place of **Rotary Motor** button.

If you want to start a new motion study, then click on the **New Motion Study** tool from the **Assembly** tab in the **Ribbon**. A new motion study tab will be added in the **MotionManager**.

Playing Motion Study

To animate the motion study, click on the **Calculate** ( ) button from the **MotionManager** tool bar. The animation of mechanism will display in the viewport. To increase the time of animation, drag the key point of assembly to the desired point; refer to Figure-62. Click on the **Calculate** button again to run the animation.

Figure-62. Keypoint of assembly

You can save the motion in movie format by selecting the **Save** ( ) button from the **MotionManager** tool bar.

BOTTOM UP APPROACH AND
TOP DOWN APPROACH

There are two ways of creating assembly in most of the CAD packages; Bottom Up approach and Top Down approach. In Bottom Up approach, all the components are created separately in Part environment and then assembled in Assembly environment by inserting them one by one.

In Top Down approach, we create all the parts in assembly environment and apply the mates on the spot. The procedure to create parts in assembly environment is given next.

Creating Parts in Assembly

• Click on the **New Part** tool from the **Insert Components** drop-down in the **Assembly** tab of **Ribbon**; refer to Figure-63. You are asked to select a face or plane on which the part is to be positioned.

Figure-63. New Part tool

- Select the desired plane from the **FeatureManager Design Tree**. The Sketching environment will be displayed; refer to Figure-64.

Figure-64. Sketching environment for assembly part

- Create desired sketch and perform operations using the tools in **Features** tab in the **Ribbon**; refer to Figure-65.

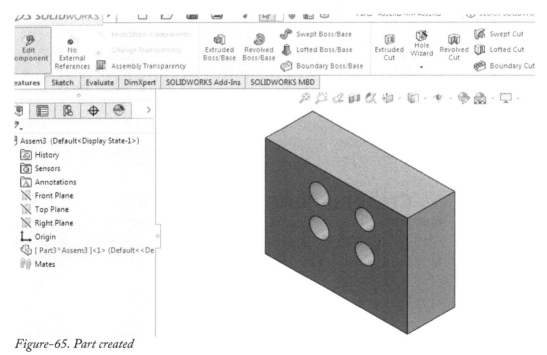

Figure-65. Part created

- Once you have created all the features of part, click on the **Exit editing component** button at the top-right in the viewport; refer to Figure-66.

Figure-66. Exit Editing Component

- Similarly, you can create other components in assembly and apply the mates as discussed earlier.

Till this point, we have used various assembly tools. In the next chapter, we will apply these tools to perform practical and work out with practice questions.

Note that for the next chapter, you need to download some part. The link to these parts is provided on writing to us at cadcamcaeworks@gmail.com

SELF ASSESSMENT

Q1. Assembly is the combination of two or more components in any manner. (T/F)

Q2. Assembly Design is a separate environment in SolidWorks to perform operations related to assembly of components. (T/F)

Q3. The first component inserted in assembly is fixed by default. (T/F)

Q4. You can reinsert a component by holding the CTRL key from keyboard while drag the desired component. (T/F)

Q5. You can make the first component transparent while applying mates in SolidWorks 2016. (T/F)

Q6. Which of the following tool can be used to manage positions of various distance and angle constraints in assembly?

a. Edit Component
b. Mate
c. Mate Controller
d. Explode View

FOR STUDENT NOTES

Chapter 8

Advanced Assembly, Practical and Practice

Topics Covered

The major topics covered in this chapter are:

- *Hole Series, Weld Bead, Chain/Belts*
- *Smart Fasteners*
- *Toolbox Utilities*
- *Magnetic Mate*
- *Asset Publisher*
- *Configurations*
- *Assembly Practical 1*
- *Assembly Practical 2*
- *Assembly and Motion Practical 3*
- *Practice Exercises.*

Note: Before starting this chapter, mail us at cadcamcaeworks@gmail.com to get the part files required to complete this chapter.

INTRODUCTION

In this chapter, you will learn about some advanced tools related to assemblies. You will also learn about handling large assemblies in SolidWorks. After that you will work on practical and practice files.

ASSEMBLY EDITING TOOLS

Most of the tools that are used for editing assembly are similar to the tools discussed for Solid model editing. The editing tools are available in the **Assembly Features** drop-down in the **Assembly CommandManager** of the **Ribbon**; refer to Figure-1. The editing tools which were not discussed earlier are discussed next.

Figure-1. Assembly Features drop-down

Creating Hole Series

The **Hole Series** tool is used to create a series of holes in the components assembled together. If you want to create single hole passing through many components in the assembly then this is the tool you should use. The procedure to use this tool is given next.

- Click on the **Hole Series** tool from the **Assembly Features** drop-down in the **Assembly CommandManager**. The **Hole Position PropertyManager** will be displayed; refer to Figure-2. Also, you will be asked to select a location for placing the new hole.

- Click at the desired location on the part. Preview of the hole will be displayed; refer to Figure-3. Click at more positions to place more holes.

Message

Sketch the location for new holes or slots or select existing holes or slots to use as the start, middle, or end of the series.

Use the Property Manager tabs to size the start, middle, and end holes or slots. Add hardware using the Smart Fasteners tab (SOLIDWORKS Toolbox required).

Hole Position

● Create new hole

○ Use existing hole(s)

Figure-2. Hole Position Proper-tyManager

Figure-3. Positiong hole series

- Every hole series has three sections; start hole, middle holes, and end hole. These sections can be edited individually. Click on the **First Part** tab in the **PropertyManager** to display options related to start hole; refer to Figure-4. Set the options as required in the **PropertyManager**.
- Similarly, click on the **Middle Parts** tab and **Last Part** tab to modify the middle and last holes in the hole series. The **PropertyManager** on selecting these tabs are shown in Figure-5.
- After specifying the desired parameters, click on the **OK** button from the **PropertyManager** to create the hole series.

Figure-5. Hole Series PropertyManager for Middle Parts and Last Part tab

Figure-4. Hole Series(First Part) PropertyManager

Creating Weld Bead in Assembly

- Click on the **Weld Bead** tool from the **Assembly Features** drop-down in the **Assembly CommandManager**. The **Weld Bead PropertyManager** will be displayed; refer to Figure-6.

Figure-6. Weld Bead PropertyManager

- Select first reference face/edge for welding.
- Click in the second selection box in the **Settings** rollout of the **PropertyManager** and select the second face/edge. Preview of the weld bead will be displayed; refer to Figure-7.

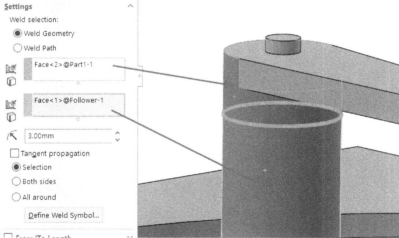

Figure-7. Faces selected for weldbead

- If you want to select the weld path in place or references then select the **Weld Path** radio button from the **Settings** rollout of the **PropertyManager** and select the path from drawing area.
- Set the other parameters of weld bead as required.

- Click on the **Define Weld Symbol** from the **PropertyManager** and set the weld symbol as required. Note that you will learn more about weld bead in chapter related to welding in this book.
- Click on the **OK** button from the **PropertyManager** to create the weld bead.

Creating Belt/Chain in Assembly

The **Belt/Chain** tool in the **Assembly Feature** drop-down is used to create representation of a real belt/chain used for motion transfer. The procedure to use this tool is given next.

- Click on the **Belt/Chain** tool from the **Assembly Features** panel in the **Assembly CommandManager** of the **Ribbon**. The **Belt/Chain PropertyManager** will be displayed; refer to Figure-8.

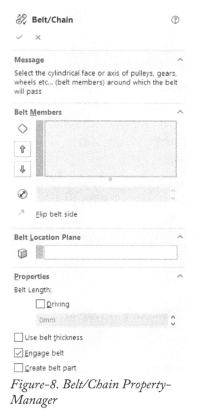

Figure-8. Belt/Chain Property-Manager

- Select the round edges of the pulley around which belt is to be created. Preview of the belt will be displayed; refer to Figure-9.

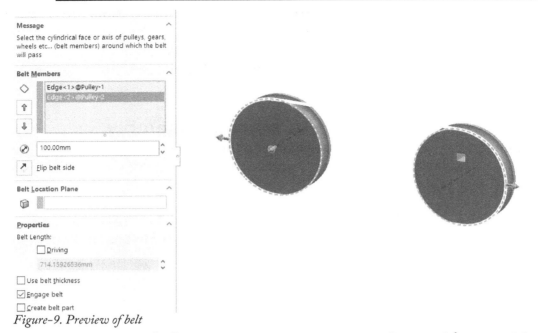

Figure-9. Preview of belt

- Select the **Engage belt** check box in the **Properties** rollout of **PropertyManager** if you want both the pulleys to behave like they are engaged by a belt. Now, if you rotate one pulley by dragging then other pulley will also rotate by same amount.
- Select the **Create belt part** check box if you want to create belt as a part for the assembly.
- Click on the **OK** button from the **PropertyManager** to create belt feature/part.

SMART FASTENERS

The **Smart Fasteners** tool is used to add fasteners to the assembly. This tool is available only when SOLIDWORKS Toolbox is installed and is active. To activate the SOLIDWORKS Toolbox, click on **SOLIDWORKS Toolbox** tool from the **SOLIDWORKS Add-Ins CommandManager** in the **Ribbon**. The procedure to use Smart Fasteners is given next.

- Click on the **Smart Fasteners** tool from the **Assembly CommandManager** in the **Ribbon**. A message box may be displayed which says smart fasteners can take longer time if there are unresolved components or there are too many holes in the assembly.
- Click on the **OK** button from the message box. The **Smart Fasteners PropertyManager** will be displayed; refer to Figure-10.

*Figure-10. Smart Fasteners
PropertyManager*

- Select the hole to which you want to apply fastener or click on the **Populate All** button from the **Selection** rollout of the **PropertyManager**. If you have selected holes manually then click on the **Add** button from the rollout. The preview of fasteners will be displayed with updated **PropertyManager**; refer to Figure-11.
- Right-click in the **Fastener** selection box in the **Series Components** rollout of the **PropertyManager**; refer to Figure-12 and select the **Change Fastener Type** option from the shortcut menu displayed. The **Smart Fastener** dialog box will be displayed; refer to Figure-13.

Figure-11. Preview of fasteners

Figure-12. Change fastener type option

Figure-13. Smart Fastener dialog box

- Select the desired fastener type and click on the **OK** button from the dialog box. The fastener in preview and **Fastener** selection box will get changed accordingly.
- By default, the **Auto size to hole diameter** and **Auto update length** check boxes are selected in the **PropertyManager** so length and size of fastener is calculated automatically. If you want to manually specify the size and length then clear the respective check boxes and specify the parameters in the **Properties** rollout of the **PropertyManager**.
- To add components like washers to the top of fastener, click in the **Add to Top Stack** drop-down. A list of options will be displayed; refer to Figure-14. Select the desired component for stacking at top. If you want to add more components in top stack then select the component again to list.

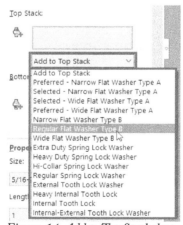

Figure-14. Add to Top Stack drop-down

- To change the parameters of the top stack component, select it from the **Top Stack** selection box and specify the parameters in the **Properties** rollout of the **PropertyManager**. You can also specify the desired parameter in the dynamic input box displayed in the preview; refer to Figure-15.

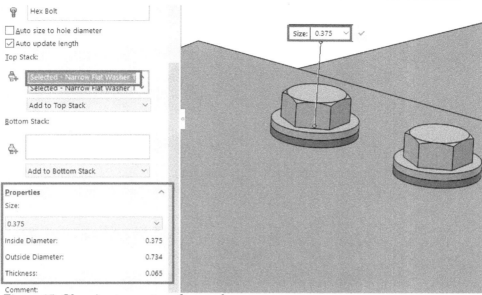

Figure-15. Changing parameters of top stack component

- Similarly, you can add components like nuts in the Bottom stack.
- After adding desired fasteners, click on the **OK** button from the **PropertyManager** to create fastener components.

TOOLBOX

There are various tools in the toolbox to increase the productivity of software in design work like Structural Steel Calculator, Beam Calculator, Cam Calculator etc. These tools are available in the **SOLIDWORKS Add-Ins CommandManager** if you have activated the **SOLIDWORKS Toolbox** button from the **CommandManager. Note that you can use most of these tools in Part environment only so you need to create a new file in Part environment to use these tools.** Various tools in the SOLIDWORKS Toolbox are discussed next.

Structural Steel Beam Calculator

The **Structural Steel Calculator** tool is used to create section of the structural steel based on the inputs provided by you. The procedure to use this tool is given next.

- Click on the **Structural Steel Beam Calculator** tool from the **SOLIDWORKS Add-Ins CommandManager**. The **Structural Steel** dialog box will be displayed; refer to Figure-16.

Figure-16. Structural Steel dialog box

- Select the desired beam standard and structural beam section from the respective drop-downs in the dialog box.
- Select the desired structural component from the list box. If you know the engineering aspects of using the current steel structure then click on the **Create** button to create the section of structural member.
- If you want to test the current section under load then click on the **Beam Calculator** button from the dialog box. The **Beam Calculator** dialog box will be displayed; refer to Figure-17.

Figure-17. Beam Calculator dialog box

- Select the desired load type from the **Load Type** list box by scrolling down in the dialog box.
- Select the parameter for which calculation is to be done from the **Type of Calculation** area of the dialog box.
- Set the value of load in the Load edit box and click on the **Solve** button. The deflection or Stress will be displayed in the respective edit box of the dialog box.
- Click on the **Done** button to exit the dialog box.

- If the deflection or stress is more than what is expected then perform the changes in the part. Otherwise, click on the **Create** button from the **Structural Steel** dialog box to create the sketch section of part.
- Click on the **Done** button from the dialog box to exit.
- Now, you can use the **Extrude Boss/Base** tool to create the beam.

Creating Grooves

The **Grooves** tool in the **SOLIDWORKS Add-Ins CommandManager** is used to create O-ring and retaining grooves on the solid. The procedure to use this tool is given next.

- Click on the **Grooves** tool from the **SOLIDWORKS Add-Ins CommandManager** in the **Ribbon**. The **Grooves** dialog box will be displayed; refer to Figure-18.

Figure-18. Grooves dialog box

- Click on the **O-Ring Grooves** tab if you want to create a O-ring groove in the solid or click on the **Retaining Ring Grooves** tab if you want to create a retaining groove in the solid.
- Select the standard for rings from the drop-down in the dialog box and then select the groove type from the drop-down below it.
- Select the size of groove from the list box. Details of the groove will be displayed in the **Property** box and preview will be displayed in the **Preview** area of the dialog box.
- Click on the solid revolve feature at desired location to create groove. A sketch will be created with a point at specified location; refer to Figure-19.
- Click on the **Create** button from the dialog box. The groove will be created; refer to Figure-20.

Figure-19. Point selected for groove feature

Figure-20. Groove created

• Click on the **Done** button from the dialog box to exit.

Creating Cams

The **Cams** tool is used to create circular and linear cams. The procedure to use this tools is given next.

• Click on the **Cams** tool from the **SOLIDWORKS Add-Ins CommandManager** in the **Ribbon**. The **Cam** dialog box will be displayed; refer to Figure-21.
• Set the desired unit in the **Units** field of the table.

Circular Cam

• Select the **Circular** option from the **Cam Type** field in the table.
• Set the desired parameters in the table.

- Click on the **Motion** tab and using **Add** button, create the motion of Cam; refer to Figure-22.

Figure-21. Cam-Circular dialog box

Figure-22. Motion setup

- Similarly, click on the **Creation** tab and set the physical parameters of the cam; refer to Figure-23.

Setup   Motion   Creation

Property	Value	
Description	Circular, Translating, 4 Motions	
Creation Method	Create new part with track removed	
Blank Outside Dia & Thickness	250	25
Near Hub Dia & Length	50	5
Far Hub Dia & Length	50	5
Blank Fillet Rad & Chamfer	1	1
Thru Hole Dia	38.1	
Track Type & Depth	Blind	20

Figure-23. Physical parameters of cam in Creation tab

- Click on the **Create** button from the dialog box to create cam; refer to Figure-24.

Figure-24. Circular cam created

- Click on the **Done** button to exit the dialog box.

Creating Linear Cam

- Select on the **Linear** option from the **Cam Type** drop-down in the table. The **Cam-Linear** dialog box will be displayed; refer to Figure-25.

- Set the parameters as discussed earlier and click on the **Create** button. The cam will be created; refer to Figure-26.

Figure-25. Cam–Linear dialog box

Figure-26. Linear cam created

- To add the current cam in the favorite list, click on the New button at the bottom of the dialog box and specify the name of the cam in the New Favorite Name dialog box displayed. Select the **Template** check box to make it as template.
- Click on the **OK** button from the dialog box to save it as favorite.
- To load previously saved cam, click on the **List** button from the dialog box and select the desired cam from the **Favorite** dialog box displayed. Click on the **Load** button from the dialog box.
- Click on the **Done** button from the dialog box to exit the tool.

Bearing Calculator

The **Bearing Calculator** tool is used to calculate the load bearing capacity of bearing and life time of bearing. You can use the tool in the same way as discussed for **Structural Steel Beam Calculator** tool.

MAGNETIC MATES

Magnetic mates are used to assemble the components automatically based on magnetic snap points defined on them. The magnetics mates are specially beneficial for piping and ducting work. It can also be used in large plant layouts where applying mates on many instances of same component becomes very time consuming. To use magnetic mates, first we need to prepare components for that. The tool used for preparing component is discussed next.

Asset Publisher

The **Asset Publisher** tool is used to create connection points on components so that they can snap to each other in assembly. The procedure to use this tool is given next.

- Open the part file which you want to use in magnetic mates like file of pipe, duct etc.
- Click on the **Asset Publisher** tool from the **Tools** menu; refer to Figure-27. The **Asset Publisher PropertyManager** will be displayed; refer to Figure-28. Also, you will be asked to select a flat face to be attached to ground plane while inserting this component in assembly.

Figure-27. Asset Publisher tool

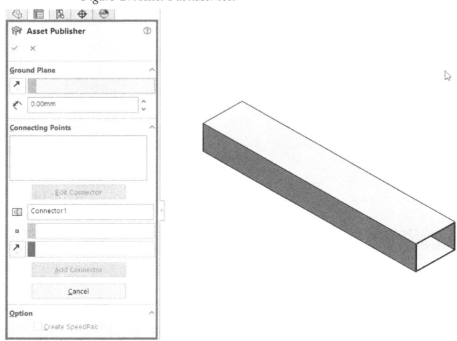

Figure-28. Asset Publisher PropertyManager with model of a duct

- Select the flat face of the component; refer to Figure-29. If you want the part to be above or below the ground plane then specify the desired distance value in the edit box in **Ground Plane** rollout of the **PropertyManager**. You can reverse the direction of placement by using the **Reverse Direction** button in the rollout.

Figure-29. Face of duct selected for ground plane

- Now, click on the edge/point on the part that you want to use as connector point and then select a face to specify direction of connection; refer to Figure-30. Note that you can cycle through the connector points by using the **[** and **]** key from keyboard.

Figure-30. Edge and face selected for connection point

- Click on the **Add Connector** button to add it in **Connecting Points** list box of **PropertyManager**.
- Repeat the previous two steps to create more connection points on the part.
- After specifying desired connection points, click on the **OK** button from the **PropertyManager**.

In the same way, you can prepare other parts for magnetic mates. Now, we have a part with connection points ready for magnetic mates. But, what if we need different lengths/sizes of same component. To enable fast conversion of size of same part, we use configurations. The procedure to create configuration is given next.

Creating Configurations

- Click on the **ConfigurationManager** tab in the **Manager** pane in the left of application window. The **ConfigurationManager** will be displayed; refer to Figure-31.

Figure-31. ConfigurationManager

- Right-click on the part name in the **ConfigurationManager**. A shortcut menu will be displayed.
- Select the **Add Configuration** option from the shortcut menu; refer to Figure-32. The **Add Configuration PropertyManager** will be displayed; refer to Figure-33.

Figure-32. Add Configuration option

Figure-33. Add Configuration PropertyManager

- Specify desired name of the configuration like "length 2k" in the **Configuration name** edit box of **PropertyManager**.
- Specify the desired description of configuration in the **Description** edit box of **PropertyManager**.
- Set the other options as required and then click on the **OK** button. Now, the newly created configuration is active.
- Change the parameters of part like length of extrusion, width in sketch etc. by using **FeatureManager Design Tree**.

- After performing changes, save the file.
- Repeat the procedure to create more configurations.
- You can check different sizes of the part by double-clicking on the respective configurations in the **ConfigurationManager**.

Note that you can create the configurations of assembly in the same way.

Creating Assembly with Magnetic Mates

- Start a new assembly as discussed earlier.
- Press **ESC** if the **Open** dialog box is displayed automatically and exit the **Begin Assembly PropertyManager** displayed automatically on starting new assembly.
- Click on the **Ground Plane** tool from the **Insert->Reference Geometry** menu; refer to Figure-34. The **Ground Plane PropertyManager** will be displayed; refer to Figure-35.

Figure-35. Ground Plane
PropertyManager

Figure-34. Ground Plane tool

- Select the plane/face that you want to use as ground plane for magnetic mate components.
- Click on the **OK** button from the **PropertyManager**.
- Click on the **Insert Components** tool from the **Assembly CommandManager** and double-click on the part (in **Open** dialog box displayed) that you have prepared for magnetic mate earlier. The part will get attached to cursor; refer to Figure-36. Note that the pink dots on the part are connection points create earlier using the **Asset Publisher** tool.
- Select the desired configuration of part from the **Configuration** drop-down in the **Insert Component PropertyManager**; refer to Figure-37.

Figure-36. Part attached to cursor with magnetic connectors

Figure-37. Configuration drop-down1

- Click on the **OK** button from the **PropertyManager**. The part will be place automatically based on the ground plane specified in part and assembly.
- Again, click on the **Insert Components** tool and double-click on the second part to be inserted. Now, if you move the part near the other part, the connections point will snap each other and purple line of connection will be displayed; refer to Figure-38.

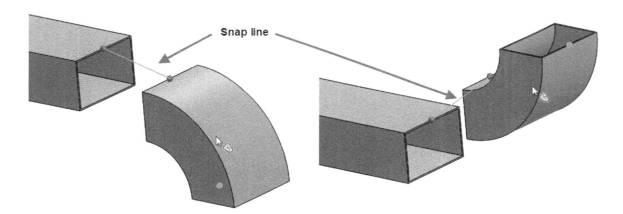

Figure-38. Autosnapping of magnetic connectors

- Click in the viewport when desired connection points are snapping. The parts will be assembled by magnetic mate; refer to Figure-39. Note that you can use the **Rotate** toolbar to rotate the part if it is not positioned as required.

Figure-39. Part connected with magnetic mate

- In the same way, you can insert the other components with connection points.

Note that you should not specify the ground plane for components when the same component is to be assembled in different orientations. Like, we have not specified ground plane for Duct Turn component in example discussed here. The part file for this example are available in the Magnetic Mates folder of Chapter 7 resources in resource kit. You can check the effect of specifying ground plane Duct turn component by yourself.

Practical 1

Assemble the parts of Fuel Injection Nozzle as shown in Figure-40. The exploded view if assembly is displayed as shown in Figure-41.

Figure-40. Assemble view of nozzle

Figure-41. Exploded view of assembly

All the part files can be downloaded from the provided link. In this practical, we will use parts of **Fuel injection nozzle** folder from downloaded files/folders in resource kit.

The steps to assemble these parts are given next.

Inserting Body

- Start the assembly environment by selecting **Assembly** button from the **New SolidWorks Document** dialog box. The **Open** dialog box will display. You are asked to insert first part.
- Double click on the Body part in the **Fuel injection nozzle** folder. The part will attach to the cursor.
- Click in the viewport to place the part.

Inserting and constraining Washer

- Click on the **Insert Components** button from the **Ribbon**. And then click on the **Browse** button from the **PropertyManager**. The **Open** dialog box will display.
- Double click on the **Washer** part in the **Fuel injection nozzle** folder.
- The part will attach to the cursor.
- Click in the viewport to place the part.
- Click on the **Mate** button from the **Ribbon**. The **Mate PropertyManager** will display.
- Select the **Coincident** button from the **PropertyManager** and select the round edges as shown in Figure-42.

Edges
selected

Figure-42. Edges to be selected

Inserting and constraining Spring

- Click on the **Insert Components** button from the **Ribbon**. And then click on the **Browse** button from the **PropertyManager**. The **Open** dialog box will display.
- Double click on the **Spring** part in the **Fuel injection nozzle** folder.
- The part will attach to the cursor.
- Click in the viewport to place the part.
- Click on the **Mate** button from the **Ribbon**. The **Mate PropertyManager** will display.
- Select the **Coincident** button from the **PropertyManager** and select the flat faces as shown in Figure-43.
- Click on the **Flip Mate Alignment** button(⚒) to align the spring properly.

Figure-43. Faces to be selected

- Click on the **OK** button from the **PropertyManager** to apply the mate.
- Click on the **View Temporary Axes** button and **View Axes** button from the **Hide/Show Items** drop-down to display axes; refer to Figure-44.

Figure-44. View Temporary Axes button

- Click on the **Coincident** button again from the **PropertyManager** and select the axes as shown in Figure-45.

Figure-45. Axes to be selected

- Click **OK** button from the **PropertyManager**. The mate will be applied.

Inserting and constraining Pre adjusting Nut

- Click on the **Insert Components** button from the **Ribbon** and then click on the **Browse** button from the **PropertyManager**. The **Open** dialog box will display.
- Double-click on the **Spring** part in the **Fuel injection nozzle** folder.
- The part will attach to the cursor.
- Click in the viewport to place the part.
- Click on the **Mate** button from the **Ribbon**. The **Mate PropertyManager** will display.
- Select the **Coincident** button from the **PropertyManager** and select the flat faces as shown in Figure-46.

Figure-46. Faces to be coincident

- Click **OK** from the **PropertyManager** to apply the mate.
- Click on the **Coincident** button again from the **PropertyManager** and select the center axes as shown in Figure-47.

Figure-47. Axes to be coincident

- Click on the **OK** button from the **PropertyManager** to apply the mate.

Inserting and Constraining other components

In the same way, you can assemble the other components of the injection nozzle; refer to Figure-40 and Figure-41.

Practical 2

Assemble the parts of handle as shown in Figure-48. The exploded view of assembly is displayed as shown in Figure-49.

Figure-48. Handle assembled

Figure-49. Exploded view of handle

All the part files can be downloaded from the provided link. In this practical, we will use parts of **Handle assembly** folder from downloaded files/folders.

The steps to assemble these parts are given next.

Inserting Main Handle

- Start the assembly environment by selecting **Assembly** button from the **New SolidWorks Document** dialog box. The **Open** dialog box will display. You will be asked to insert first part.
- Double-click on the **Main Handle** part in the **Handle assembly** folder. The part will attach to the cursor.
- Click in the viewport to place the part.

Inserting and constraining Handle

- Click on the **Insert Components** button from the **Ribbon**. And then click on the **Browse** button from the **PropertyManager**. The **Open** dialog box will display.

- Double click on the **Handle** part in the **Handle assembly** folder.
- The part will attach to the cursor.
- Click in the viewport to place the part.
- Click on the **Mate** button from the **Ribbon**. The **Mate PropertyManager** will display.
- Select the **Concentric** button from the **PropertyManager** and select the round faces as shown in Figure-50.

Faces to be selected

Figure-50. Faces selected for concentric mate

- Click **OK** from the **PropertyManager** to apply the mate.
- Display the planes by selecting the **View Planes** button from the **Hide/Show Items** drop-down in the **Heads-up View** tool bar; refer to Figure-51.

Figure-51. View Planes button

- Click on the **Coincident** button from the **Mate PropertyManager** and select the plane and edge as shown in Figure-52.

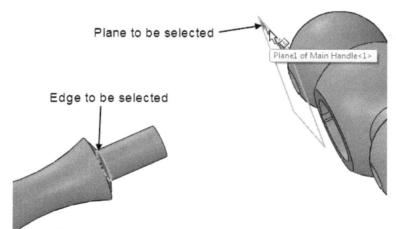

Plane to be selected

Plane1 of Main Handle<1>

Edge to be selected

Figure-52. Selection for coincident mate

- Click on the **OK** button from the **PropertyManager** to apply the mate.

The assembled handle will displayed as shown in Figure-53.

Figure-53. Assembled handle

Similarly, assemble the other handle so that the model displays as shown in Figure-48.

Practical 3

Assemble the parts of Epicyclic Gear Mechanism as shown in Figure-54. The exploded view if assembly is displayed as shown in Figure-55. Also, apply the simulation to mechanism as shown in video file provided in the resources.

Figure-54. Epicyclic gear mechanism assembly

Figure-55. Exploded view of epicyclic gear mechanism

All the part files can be downloaded from the provided link. In this practical, we will use parts of **Epicyclic Gear Mechanism** folder from downloaded files/folders.

The steps to assemble these parts are given next.

Inserting Static Ring

- Start the assembly environment by selecting Assembly button from the **New SolidWorks Document** dialog box. The **Open** dialog box will display. You will be asked to insert first part.
- Double-click on the Static Ring part in the **Epicyclic Gear Mechanism** folder. The part will attach to the cursor.
- Click in the viewport to place the part. The part will be fixed at the specified position.

Inserting and constraining Washer

- Click on the **Insert Components** button from the **Ribbon**. And then click on the **Browse** button from the **PropertyManager**. The **Open** dialog box will display.
- Double click on the **Eccentric Input shaft** part in the **Epicyclic Gear Mechanism** folder.
- The part will attach to the cursor.
- Click anywhere in the viewport to place the component.
- Select the round face of **Input shaft** and **Static ring**, and select the **Concentric** button from the **Mates** toolbar displayed; refer to Figure-56.

Figure-56. Applying concentric mate

- Select the flat face of Input shaft and Static ring as shown in and apply the **Coincident** mate from the **Mates** toolbar.

Inserting and constraining Eccentric static bearing

- Click on the **Insert Components** button from the **Ribbon**. And then click on the **Browse** button from the **PropertyManager**. The **Open** dialog box will display.
- Double click on the **Static Eccentric Bearing** part in the **Epicyclic Gear Mechanism** folder.
- The part will attach to the cursor.
- Click anywhere in the viewport to place the component.
- Select the round face of **Input shaft** and **Static Eccentric Bearing**, and select the **Concentric** button from the **Mates** toolbar displayed; refer to Figure-57.

Figure-57. Applying concentric mate on bearing

- Similarly, make the flat faces of the bearing and shaft coincident; refer to Figure-58.

Figure-58. Making flat faces coincident

Inserting and constraining Planet gear

- Click on the **Insert Components** button from the **Ribbon**. And then click on the **Browse** button from the **PropertyManager**. The **Open** dialog box will display.
- Double click on the **Planet Gear** part in the **Epicyclic Gear Mechanism** folder.
- The part will attach to the cursor.
- Click anywhere in the viewport to place the component.
- Select the round face of **Input shaft** and **Plant Gear**, and select the **Concentric** button from the **Mates** toolbar displayed; refer to Figure-59.

Figure-59. Applying concentric mate on planet gear

- Similarly, applying the coincident mate on the flat face of gear and Input shaft; refer to Figure-60.

Figure-60. Making flat faces of gear and shaft coincident

Inserting and constraining other components

In the same way, insert and constrain the other components refer to the figures given next for reference.

Figure-61. Placing arm pin

Figure-62. Placing the rod

Figure-63. Placing output crank

Figure-64. Placing the output shaft bearing

Applying relation between gears

- After placing all the components and applying the constraints, click on the **Mate** button from the **Ribbon**. The **Mate PropertyManager** will be displayed. Select the **Gearmate** button from the **Mechanical Mates** rollout in the **PropertyManager**; refer to Figure-65.

Figure-65. Gear mate

- Select the two edges of gear and static ring; refer to Figure-66.

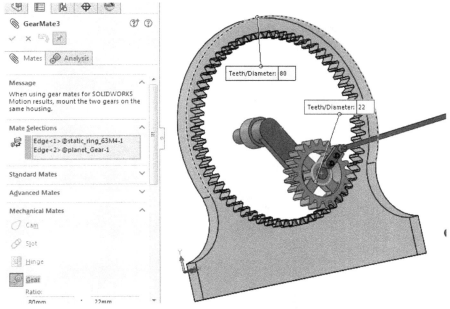

Figure-66. Edges selected for gear mate

- Specify the gear ratio as **80:22** in the **Mechanical Mates** rollout in the **PropertyManager** and click on the **OK** button from it twice.

Creating Motion Study

- Click on the **Motion Study 1** tab from the bottom bar. The Motion Study interface will be displayed; refer to Figure-67.

Figure-67. Motion study for gear

- Click on the **Motor** button from the Motion Study interface.
- Select the face shown in Figure-68. Specify the RPM as **10** and click on the **OK** button from the **Motor PropertyManager**.

Figure-68. Face selected for motor

- Select the **Basic Motion** option from the drop-down; refer to Figure-69 and then click on the **Calculate** button to check the motion.

Figure-69. Basic Motion option

PRACTICE 1

Assemble the model as shown in Figure-70. The exploded view is given in Figure-71.

Figure-70. Bottom assembly

Figure-71. Exploded view of bottom assembly

Note that you need to import nuts and bolts from the toolbox provided by SolidWorks; refer to Figure-72. The toolbox is the standard library of components provided by SolidWorks. This library contains nuts, bolts, bearings, transmission parts, washer and a lot more with inch as well as mm specifications.

Figure-72. Toolbox

PRACTICE 2

Assemble the model as shown in Figure-73. The exploded view is given in Figure-74.

Figure-73. Top Assembly isometric view

Figure-74. Exploded top assembly

PRACTICE 3

Assemble the Top Assembly and Bottom Assembly created in Practice 1 and Practice 2 to for the assembly as shown in Figure-75.

Note that in this practice, you will insert the sub-assemblies in place of inserting single parts.

Figure-75. Assembly for practice 3

To get more parts for practice, write us at cadcamcaeworks@gmail.com

FOR STUDENT NOTES

FOR STUDENT NOTES

Chapter 9

Surfacing and Practice

Topics Covered

The major topics covered in this chapter are:

- **Surfacing Introduction.**
- **Surfacing tools similar to solid creation tools.**
- **Special tools for surfacing.**
- **Surface editing tools.**
- **Surface to solid conversion.**
- **Practical and Practice.**

Extruded Revolved Swept Lofted Boundary Filled Freeform   Planar Surface   Surface Fillet   Delete Face   Knit   Thicken   Referenc... Curves
Surface Surface Surface Surface Surface Surface   Offset Surface   Flatten   Replace Face   Surface   Thickened Cut
Ruled Surface   Cut With Surface

Features   Sketch   **Surfaces**   Sheet Metal   Weldments   Evaluate   DimXpert   SOLIDWORKS Add-Ins   SOLIDWORKS MBD

SURFACING

Surfacing is a separate world in the field of CAD. The complicated shapes which are difficult for solid modeling are most of the time easy for surfacing. Basic tools of surfacing are very similar to the solid creation tools discussed earlier like, extrude, revolve, sweep and so on. But there are some other tools that allow to modify the part shape freely in 3D. To start surfacing in SolidWorks, click on the **Surfaces** tab in the **Ribbon** in Part environment. If the tab is not available by default, then right-click on any of the tab in the **Ribbon** and select the **Surfaces** option from the menu. The interface will display as shown in Figure-1. All the tools are used in SolidWorks for surfacing are explained as follow:

Figure-1. Surfacing interface

SURFACING TOOLS SIMILAR TO SOLID CREATION TOOLS

The list of these tools is given next.

- **Extruded Surface**   similar to   **Extruded Boss/Base**
- **Revolved Surface**   similar to   **Revolved Boss/Base**
- **Swept Surface**   similar to   **Swept Boss/Base**
- **Lofted Surface**   similar to   **Lofted Boss/Base**
- **Boundary Surface**   similar to   **Boundary Boss/Base**

These tools are one by one explained next.

Extruded Surface

The **Extruded Surface** tool is used to extrude a close or open sketch to the specified height to form a surface. The steps to use this tool are given next.

- Click on the **Extruded Surface** tool from the **Ribbon.** You will be asked to select a plane to draw sketch.
- Draw open or close sketch of the surface on desired plane.
- Exit the sketch. Preview of the surface will display; refer to Figure-2.

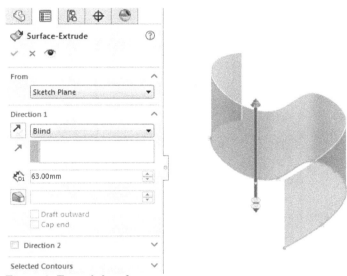

Figure-2. Extruded surface

- If you draw a close sketch then you can select the **Cap end** check box to close the end edges.
- The other options are similar to **Extruded Boss/Base** tool.
- Click on the **OK** button from the **PropertyManager** to create the extruded surface.

Revolved Surface

The **Revolved Surface** tool is used to revolve a close or open sketch to the specified angle with respect to selected reference to form a surface. The steps to use this tool are given next.

- Click on the **Revolved Surface** tool from the **Ribbon.** You will be asked to select a plane to draw sketch.
- Draw open or close sketch of the surface on desired plane. Make sure that you create a center line for revolving the sketch.
- Exit the sketch. Preview of the surface will display; refer to Figure-3. Options in the **PropertyManager** have been discussed already.

Figure-3. Revolved surface

- Click on the **OK** button from the **PropertyManager** to create the revolved surface.

Swept Surface

The **Swept Surface** tool is used to sweep a section along the selected path to form a surface. The steps to use this tool are given next.

- Click on the **Swept Surface** tool from the **Ribbon**. **Surface-Sweep PropertyManager** will display.
- Select the open or close section sketch.
- Select the open or close sketch for the path. Preview of the swept surface will display; refer to Figure-4.

Figure-4. Swept surface

- The options in the **PropertyManager** are same as for **Swept Boss/Base** tool.
- Click on the **OK** button from the **PropertyManager** to create the surface.

Note that for creating **Swept Boss/Base** feature as well as for **Swept surface,** you can select edge of an existing model as a path if required.

Lofted Surface

The **Lofted Surface** tool is used to join two open/close sections to form surface. The steps to use this tool are given next.

- Click on the **Lofted Surface** tool from the **Ribbon**. **Surface-Loft PropertyManager** will display.
- Select the first open/close section sketch.
- Select the second open/close section sketch. Preview of the lofted surface will display; refer to Figure-5.

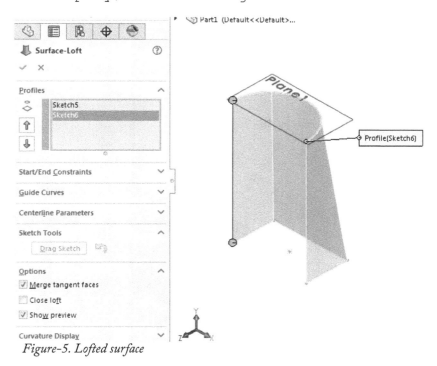

Figure-5. Lofted surface

- Click **OK** button to create the surface.

Note that if you are selecting first section as open sketch then the second section should also be open sketch. Same with the close sketches.

Boundary Surface

The **Boundary Surface** tool is used to join two open/close sections to form surface. The steps to use this tool are given next. This tool works in the same way as the lofted surface.

Note that using the Lofted Surface tool and Boundary Surface tool, you can join two or more surfaces, edges of solids and face of solids. Refer to Figure-6 and Figure-7.

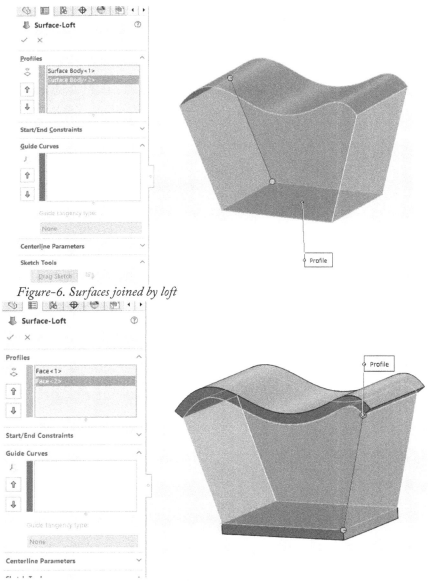

Figure-6. Surfaces joined by loft

Figure-7. Faces of solids joined by loft

Filled Surface

The **Filled Surface** tool is use to fill gap by selecting the close boundary. This close boundary can be made by creating sketches or it can be formed by intersection of solids/surfaces. The steps to use this tool are given next.

- Click on the **Filled Surface** tool from the **Ribbon**. The **Fill Surface PropertyManager** will display; refer to Figure-8.

Figure-8. Fill Surface PropertyManager

- Select the boundary of area that you want to fill using the surface. Preview of the surface will display; refer to Figure-9.

Boundary lines are created used two different 3D sketches to close the surface

Figure-9. Filled surface

This tool is the most used one for closing surfaces. Using this tool, you can also close circular holes in surfaces/solids.

Freeform

The **Freeform** tool is used to freely deform solid faces/surfaces. The steps to use this tool are given next.

- Click on the **Freeform** tool from the **Ribbon**. The **Freeform PropertyManager** will display; refer to Figure-10.

Figure-10. Freeform PropertyManager

- Select the surface/face you want to deform. Mesh on curves will display on the surface.
- Click on the **Add Curves** button and click at the desired position over the surface to set the highlighted line as control curve. You can select more than one curve for controlling the surface shape.
- Click again on the **Add Curves** button to exit the selection mode.
- Click on the **Add Points** button to add control points on the control curve.
- Click again on the **Add Points** button or Right-click to exit selection mode.
- Drag the point you have created earlier to change the shape of the surface/face. Refer to Figure-11.

Figure-11. Freeforming surface

- Click on the **OK** button from the **PropertyManager** to apply the modification.

SPECIAL SURFACING TOOLS

Earlier, you have learned the tools that were similar to the Solid creation tools. Now, we will discuss about the special tools provided in SolidWorks to create surfaces. These tools are discussed next.

Planar Surface

The **Planar Surface** tool is used to create a surface joining two or more edges or a sketch that is in the same plane. The steps to use this tool are given next.

- Click on the **Planar Surface** tool from the **Ribbon**. The **Planar Surface PropertyManager** will display; refer to Figure-12.

Figure-12. Planar Surface PropertyManager

- Select the edges that form a planar surface. The preview will be displayed; refer to Figure-13.

Figure-13. Preview of planar surface

- Click on the **OK** button from the **PropertyManager** to create the surface.

Offset Surface

The **Offset Surface** tool is used to create a surface at an offset distance from the selected face/surface. The steps to create offset surface are given next.

- Click on the **Offset Surface** tool from the **Ribbon**. The **Offset Surface PropertyManager** will display; refer to Figure-14.
- Select a surface/face by which you want to create the offset surface. Preview of the surface will be displayed; refer to Figure-15.

Figure-14. Offset Surface PropertyManager

Figure-15. Preview of offset surface

- Specify the desired distance in the spinner and click on the **OK** button to create the offset surface.

Ruled Surface

The **Ruled Surface** tool is used to create a combination of surface adjoining to each other. This type of surface becomes very important while creating parting surface for molding/casting. The steps to use this tool are given next.

- Click on the **Ruled Surface** tool from the **Ribbon**. The **Ruled Surface PropertyManager** will display; refer to Figure-16.
- Select the edges using which you want to create the ruled surface. Preview of surface will display; refer to Figure-17.

- Increase the length of surfaces by using the spinner in the **Distance/ Direction** rollout.

Figure-17. Preview of ruled surface

Figure-16. Ruled Surface Property-Manager

- If you select the **Normal to Surface** radio button, then the surfaces will be displayed as shown in Figure-18.

Figure-18. Preview of surface on selecting Normal to Surface radio button

- If you want to create surface tapered to the selected vector, then select the **Tapered to Vector** radio button and then select the edge to specify vector. Also, set the angle in the **Angle** spinner. Preview of the surface will be displayed; refer to Figure-19.

Figure-19. Preview of surface on selecting Tapered to Vector radio button

- Select the **Perpendicular to Vector** radio button to create the surfaces perpendicular to selected vector. Refer to Figure-20.

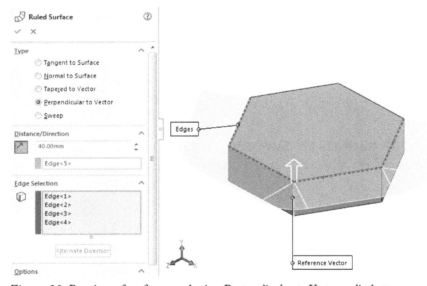

Figure-20. Preview of surface on selecting Perpendicular to Vector radio button

- Select the **Sweep** radio button to sweep the surface along the selected vector. Preview will be displayed; refer to Figure-21.
- After selecting the desired option, click on the **OK** button to create the surfaces.

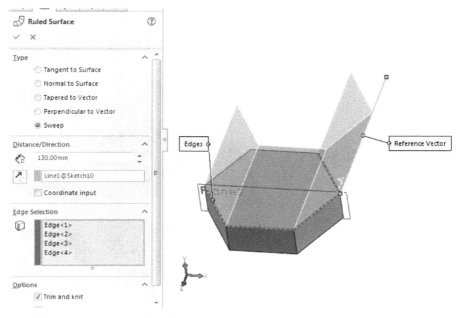

Figure-21. Preview of surface on selecting Sweep radio button

Surface Flatten tool

The **Surface Flatten** tool was added in SolidWorks 2015. The **Surface Flatten** tool as the name suggests is used to make a flat surface of various interconnected surfaces. Note that before using this tool, there must be surfaces in the viewport. The procedure to use this tool is given next.

*   Click on the **Surface Flatten** tool from the **Ribbon**. The **Flatten PropertyManager** will be displayed; refer to Figure-22.
*   Select the surface/surfaces that you want to flatten; refer to Figure-23.

Figure-23. Surface selected

Figure-22. Flatten Property-Manager

*   Click in the **Vertex/Point** collector to select the fixed reference.

- Select the corner vertex of the surface. Preview of the surface will be displayed; refer to Figure-24.

Figure-24. Preview of flattened surface

- If you want to apply relief in the surface, then select the **Relief Cuts** check box and click in the selection box in **Relief Cuts** rollout. You will be asked to select a curve.
- Click on the curve that you want to use for relief cut; refer to Figure-25.

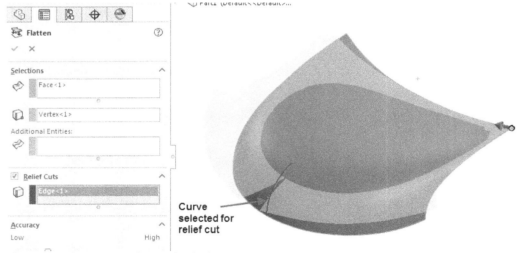

Curve
selected for
relief cut

Figure-25. Flattened surface with relief cut

- Click on the **OK** button from the **PropertyManager** to create the flattened surface. Note that the parent surface will not be deleted by this operation.

SURFACE EDITING TOOLS

The tools in this category are used to edit the surfaces. For example, trimming the surface, deleting some portion of the surface, knitting two surfaces for applying fillet. These tools are explained next.

Delete Face

The **Delete Face** tool is used to remove a face/surface from the model. The procedure to use this tool are given next.

- Click on the **Delete Face** tool from the **Ribbon**. The **Delete Face PropertyManager** will display as shown in Figure-26.
- Select the face that you want to remove.

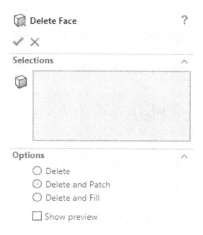

Figure-26. Delete Face PropertyManager

- Select the desired radio button, if you want to patch or fill the gap created by deleting the surface.
- Click on the **OK** button from the **PropertyManager**. The face will be deleted. Note that using this tool, you cannot delete a single face surface body.

Replace Face

The **Replace Face** tool is used to extend the selected face to the replacement face. The procedure to use this is given next.

- Click on the **Replace Face** tool. The **Replace Face PropertyManager** will display; refer to Figure-27.

Figure-27. Replace Face PropertyManager

- Select the face/faces that you want to replace.
- Click in the next selection box in the **PropertyManager** and select the surface by which you want to replace the face.

- Click on the **OK** button to apply the replacement; refer to Figure-28.

Figure-28. Output of replace face

Extend Surface

The **Extend Surface** tool is used to extend the selected surface by specified value. The procedure to use this tool is given next.

- Click on the **Extend Surface** tool. The **Extend Surface PropertyManager** will display; refer to Figure-29.
- Select the edge of the surface that you want to extend. Preview of extension will display; refer to Figure-30.

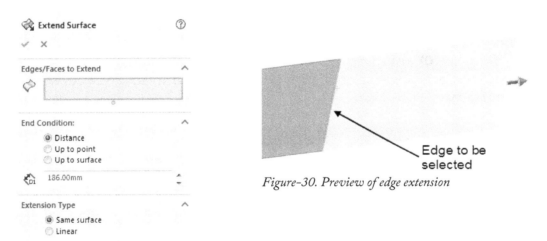

Figure-30. Preview of edge extension

Figure-29. Extend Surface PropertyManager

- Specify the length of extension in the spinner. If you want to extend the surface to a selected point/surface then select the respective radio button and then the reference. The surface will extend to that reference.

Trim Surface

The **Trim Surface** tool is used to remove desired portion of the surface by using sketch/other surface. The procedure to use this tool is given next.

- Click on the **Trim Surface** tool. The **Trim Surface PropertyManager** will display; refer to Figure-31.

- Select the sketch or surface that you want to use as trimming tool. You are asked to select the portion that you want to keep; refer to Figure-32.

Figure-32. Preview of trimmed surface

Figure-31. Trim Surface PropertyManager

- Select the portion that you want to keep. The surface on the other side of trimming surface or sketch will be removed.

Untrim Surface

The **Untrim** tool is used to undo the trimmed surfaces. The following steps explain the procedure to untrim the surface.

- Select the **Untrim Surface** tool from the **Ribbon**. The **Untrim Surface PropertyManager** will display; refer to Figure-33.

Figure-33. Untrim Surface PropertyManager

- Select the surface on which the trimming operations performed.
- Click on the **OK** button from the **PropertyManager**. The surface will be untrimmed and all the gaps will be patched.

Knit Surface

The **Knit Surface** tool is used to combine two or more surfaces at their common edges. If the surface form a close boundary then this tool turn the surfaces in to a solid. The steps to use this tool are given next.

- Click on the **Knit Surface** tool from the **Ribbon**. The **Knit Surface PropertyManager** will display; refer to Figure-34.

Figure-34. Knit Surface PropertyManager

- Select the surfaces that you want to knit with each other.
- Click **OK** button. The surfaces will be combined into a single entity. All the knit surfaces are added in the **Surface Bodies** folder; refer to Figure-35.

Figure-35. Knitted Surfaces

Note that the knitted surfaces are used as parting surfacing in molding and casting. Also, if the surfaces form a closed boundary then select the **Create solid** check box to create solid from surface.

Thicken

The **Thicken** tool is used to add thickness to the surface. The procedure to use this tool is given next.

- Click on the **Thicken** tool from the **Ribbon**. The **Thicken PropertyManager** will display; refer to Figure-36.

Figure-36. Thicken PropertyManager

- Select the surface to which you want to add thickness. The preview of thickened surface will display; refer to Figure-37.

Figure-37. Preview of thickened surface

- Specify the desired thickness in the spinner and change the side of thickness application by selecting the desired button.

Thickened Cut

The **Thickened Cut** tool is used to remove material by thickening the surface. This tool works in the same way as **Thicken** tool but it removes material in place of adding.

Cut with Surface

The **Cut with Surface** tool is used to cut solids by using the surface. The steps to use this tool are given next.

- Click on the **Cut with Surface** tool from the **Ribbon**. The **SurfaceCut PropertyManager** will display; refer to Figure-38.

Figure-38. SurfaceCut PropertyManager

- Select the surface which you want to use for cutting the solids.
- The solids in the direction of arrow will be removed. Click on the arrow to flip the direction.
- Click on the **OK** button to cut the solids; refer to Figure-39.

Figure-39. Solid cut by surface

We have covered all the important tools that are used for surfacing in SolidWorks. Now, we will practice on some models to apply these tools.

Practical 1

Create the model of helmet glass as shown in Figure-40. The dimensions of the model are given in Figure-41.

Figure-40. Practical1 model

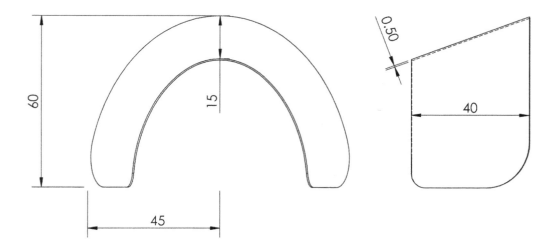

Figure-41. Practical1 drawing

The model displayed is having very low thickness and its having complex 3D shape. So, it is a good idea to use surfacing in this case.

We can create this model by lofted surface easily. For that we need to have two sketches.

Creating first sketch

- Click on the **Sketch** tab and select the **Sketch** button. You are asked to select sketching plane.
- Click on the **Top** plane. The sketching environment will display.
- Click on the **Ellipse** tool from the **Ribbon** and draw an ellipse as shown in Figure-42.
- Draw center line passing through coordinate system and trim the bottom portion of the ellipse; refer to Figure-43.

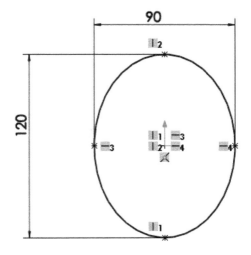

Figure-42. Ellipse to be drawn

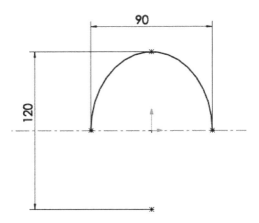

Figure-43. Trimmed ellipse

- Click on the **Exit Sketch** button to exit the sketch.

Creating second sketch

- Click on the **Plane** tool from **Reference Geometry** drop-down in **Surfaces** tab of the **Ribbon** and create a plane at an offset distance of **40** above the **Top** plane; refer to Figure-44.

Figure-44. Offset plane to be created

- Click on the **Sketch** button from the **Sketch** tab and select the **Top** plane as sketching plane. Press **CTRL+8** if sketching plane is not parallel.
- Click on the **Offset Entities** tool from the **Ribbon** and selected the earlier created sketch.
- Click on the **Reverse** check box and specify the value as **15** in the spinner; refer to Figure-45.

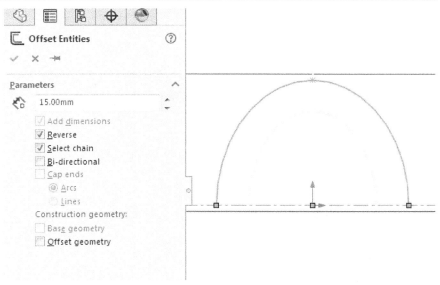

Figure-45. Preview of offset entities

- Click on the **OK** button to create the offset curve.
- Click on the **Exit Sketch** button.

Creating lofted surface

- Click on the **Lofted Surface** tool from **Surfaces** tab of the **Ribbon**. The **Surface-Loft PropertyManager** will display.
- Select the two sketches one by one. Preview of the surface will display; refer to Figure-46.

Figure-46. Preview of lofted surface

- Click on the **OK** button to create the surface.
- To hide the plane, select it and right-click on it. Select the **Hide** button from the shortcut menu box.

Thickening surface and applying fillet

- Click on the **Thicken** tool from the Ribbon and select the surface. Preview of thickened surface will display.
- Enter the thickness value as **0.5**.
- Click **OK** to create the solid.

- Click on the **Fillet** tool and apply suitable fillets at the small edges on the corners. Refer to Figure-40.

Practical 2

Create the model of flower vase as shown in Figure-47. The dimensions of the model are given in Figure-48.

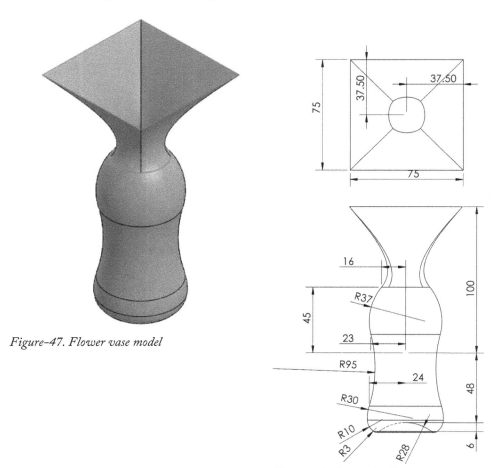

Figure-47. Flower vase model

Figure-48. Practical 2 Drawing

The model displayed is having no thickness and its having complex 3D shape. So, we will be using surfacing to create this model.

We can create this model in two steps. In step 1, we will create a revolved surface and in step 2 we will connect the revolved surface to a rectangle by using **Lofted Surface** tool. The procedure to create this surface model is given next.

Creating Revolved Surface

- Click on the **Sketch** tab and select the **Sketch** button. You are asked to select sketching plane.
- Click on the **FRONT** plane. The sketching environment will display.
- Using the **3 Point Arc** tool create the sketch as shown in Figure-49. Make sure you create a centerline as shown in the figure.

Figure-49. Sketch for revolved surface

- Click on the **Exit Sketch** button and Select the **Revolved Surface** tool from the **Surfaces** tab in the **Ribbon**.
- Select the sketch created recently. Preview of the revolved surface will be displayed; refer to Figure-50.

Figure-50. Preview of revolved surface

- Click on the **OK** button from the **PropertyManager** displayed to create the surface.

Creating Lofted Surface

- Click on the **Plane** tool from the **Reference Geometry** drop-down in the **Ribbon** and create an offset plane at distance of **100** from the TOP plane; refer to Figure-51.

- Create a rectangle on the selected plane; refer to Figure-52.

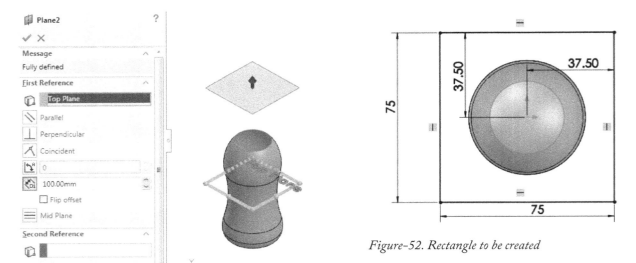

Figure-51. *Offset plane created*

Figure-52. *Rectangle to be created*

- Click on the **Lofted Surface** tool from the **Ribbon**. The **Surface-Loft PropertyManager** will be displayed.
- Select the recently created sketch and edge of the surface. Preview of the lofted surface will be displayed; refer to Figure-53.

Figure-53. *Preview of the lofted surface*

- Make sure you set the same values as highlighted in red boxes in the Figure-53. Click on the **OK** button from the **PropertyManager** to create the feature.

PRACTICE 1

Create the surface model of tank as shown in Figure-54. The dimensions of the model are given in Figure-55.

Figure-54. Practice1 model

Figure-55. Practice1 drawing

PRACTICE 2

Create the surface model of car bumper as shown in Figure-56. The dimensions of the model are given in Figure-57. **Assume the missing dimensions.**

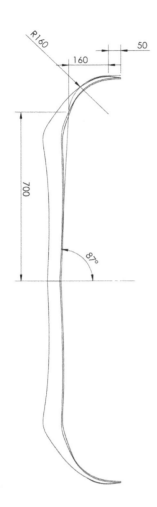

Figure-57. car bumper

Figure-56. Practice 2 model

SELF ASSESSMENT

Q1. We can create an extruded surface by using an open section. (T/F)

Q2. We do not need any centerline for creating revolved surface. (T/F)

Q3. We can create a swept surface by using open section and closed path. (T/F)

Q4. The **Freeform** tool can not deform face os a solid body.(T/F)

Q5. The **Ruled Surface** tool is used to create a combination of surfaces adjoining to each other. (T/F)

Q6. Which of the following tool can be used to create solid from the surface?

a. **Knit Surface**
b. **Untrim Surface**
c. **Trim Surface**
d. **Extend Surface**

Q7. Which of the following tool is used to remove desired portion of the surface by using sketch/other surface.

a. **Thickened Cut**
b. **Trim Surface**
c. **Cut with Surface**
d. **Delete Face**

Q8. The...........tool is used to extend the selected face to the replacement face.

Q9. The..........tool is used to undo the trimmed surfaces.

Q10. The.............tool is used to combine two or more surfaces at their common edges.

FOR STUDENT NOTES

Chapter 10

Drawing and Practice

Topics Covered

The major topics covered in this chapter are:

- *Drawing Introduction.*
- *Drawing Sheet Selection.*
- *Adding Views to sheet.*
- *Annotating Views.*
- *Exploded View and Bill of Material.*
- *Balloons and Title Block*
- *Practice.*

INTRODUCTION

Drawing is the engineering representation of a model on the paper. For manufacturing a model in real world, we need some means by which we can tell the manufacturer what to manufacture. For this purpose, we create drawings from the models. These drawings have information like dimensions, material, tolerances, objective, precautions and so on. In SolidWorks, we create drawings by using the Drawing environment. To activate this environment, click on the **New** button from the **Menu Bar**. The **New SolidWorks Document** dialog box will display. Double-click on the **Drawing** button; refer to Figure-1. The Drawing environment will open.

Figure-1. Drawing button to be selected

DRAWING SHEET SELECTION

On starting the Drawing environment, the interface will display as shown in Figure-2. Also, **Sheet Format/Size** dialog box will display. Using this dialog box, you can set the size of sheet for placing the drawing views.

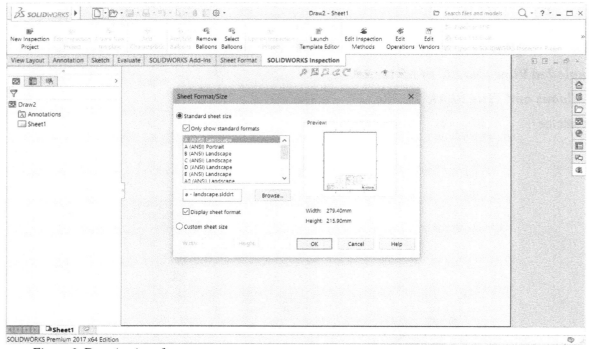

Figure-2. Drawing interface

Select the desired size from the list of sheet sizes and click on the **OK** button from the list. The **Model View PropertyManager** will become active automatically. Click on the **Cancel** button from the **PropertyManager** (We will discuss about it later in this chapter). If you want to change the sheet size, follow the steps given below.

• Right-click on the **Sheet** tab from the bottom bar displayed below drawing area. Shortcut menu will display; refer to Figure-3.

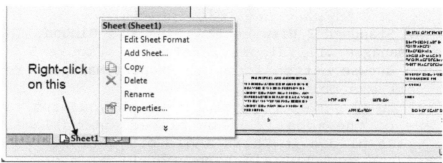

Figure-3. Shortcut menu displayed

• Select the **Properties** button from the menu. The **Sheet Properties** dialog box will display; refer to Figure-4.

Figure-4. Sheet Properties dialog box

• Select the desired sheet size from the list in this dialog box. You can also set the scale for view and projection type (First angle or Third angle) by using the options in this dialog box.
• You can set you custom dimensions by selecting the **Custom sheet size** radio button. Click on the **OK** button to accept the sheet size.

After setting the sheet size, the next step is to add various views of the model in the sheet.

ADDING VIEWS TO SHEET

The tools to add views are available in the **View Layout** tab of the **Ribbon**. These tools are explained next.

Standard 3 View

The **Standard 3 View** tool is used to add 3 standard view (Top view, Front view, and Right view) to the sheet. To add these views, follow the steps given next.

*   Click on the **Standard 3 View** button from the **Ribbon**. The **Standard 3 View PropertyManager** will display; refer to Figure-5.
*   Click on the **Browse** button from the **PropertyManager**. You are asked to select part/assembly model.
*   Double-click on the file of model for which you want to create the views. Views of the model will be placed automatically. Figure-6 shows the views placed in Third angle projection.
*   You can drag the view by selecting any curve of model to place at your desired location. The other views will be shifted accordingly; refer to Figure-7.

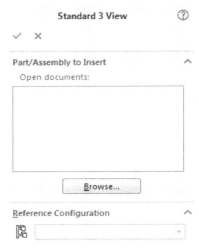

Figure-5. Standard 3 View Property-Manager

Figure-6. 3 Views placed automatically

Figure-7. Moving views

To change the properties of any of the view placed, select it from the drawing. The **Drawing View PropertyManager** will display. If you select the main view, then the **PropertyManager** will display as shown in Figure-8. If you select any projected view then the **PropertyManager** will display as shown in Figure-9.

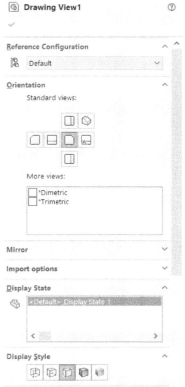

Figure-8. Drawing View Property-Manager on selecting main view

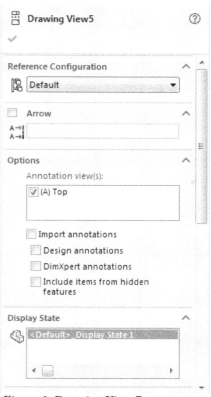

Figure-9. Drawing View Property-Manager on selecting projected view

The options in the **PropertyManager** are discussed next.

Reference Configuration rollout

Select the desired configuration of part for which you want to create the drawing views from the drop-down in this rollout.

Orientation rollout

Select the desired button to change the view orientation of the model. You can display dimetric or trimetric orientation by selecting the respective check box.

Mirror rollout

This rollout was added in Solidworks 2017. The rollout is available for main view only. Select the **Mirror view** check box in this rollout and select the **Horizontal** or **Vertical** radio button to mirror the selected view horizontally or vertically.

Import Options rollout

Options in this rollout are used to import information from the Part environment. If you want to use dimensions or other annotations created in Part environment, then select the **Import annotations** check box. The other check boxes below it will become active. Similarly, you can select other check boxes to include design annotations, DimXpert annotations and hidden features from Part environment.

Display Style rollout

Options in this rollout are used to change the display of model view. If you want to display hidden lines then select the **Hidden Lines Visible** button from the rollout. Similarly, you can select the other buttons as per your requirement.

Scale rollout

Options in this rollout are used to change the size of view by using the scale value. To set the desired scale, click on the **Use custom scale** radio button from the rollout and enter the desired ratio in the edit box below it or select the desired ratio from the drop-down below the radio buttons.

Similarly, you can set the thread display quality and dimension type from the respective rollouts.

If you have selected a projected view for changing properties, then you can add the projection arrows by selecting the **Arrow** check box from the **PropertyManager**; refer to Figure-10. Enter the name of projection arrows in the edit box displayed in the rollout.

Figure-10. Arrow for projection

Now, we will learn to place the main view in the sheet. If you do not want to use the Standard 3 Views then the next tool is the first step for drawing. **Note that if you want to delete any view/views from the sheet, then select it and press DEL key from keyboard.**

Model View

The **Model View** tool is used to add main view in the drawing. All the other views in the drawing are projection of it and are related to it. The main view can be of any orientation like, front, left, top and so on. To use this tool, follow the steps given next.

*   Click on the **Model View** button from the **Ribbon**. The **Model View PropertyManager** will display.
*   Click on the **Browse** button from the **PropertyManager**. The **Open** dialog box will display.
*   Double click on the part/assembly model file to generate the view. The first view of model will get attached to the cursor; refer to Figure-11.
*   Select the desired orientation and other parameters in the **PropertyManager,** and click in the sheet to place the view.

Figure-11. View attached to cursor

*   Projection mode of the tool will start and you will be prompted to click in the sheet to place the projection view.
*   Click to place the projection view.
*   Press **ESC** key to exit the tool.

Projected View

The **Projected View** tool is used to create projected view of the selected view in the sheet. The procedure to use this tool is given next.

*   Click on the **Projected View** tool from the **Ribbon**. The **Projected View PropertyManager** will be displayed.
*   Click in the sheet to place the projection of view available in the sheet. Note that to use this tool you must have a view in the sheet.
*   If you have more than one view in the sheet then select the view whose projection is to be generated and then click to place the projection; refer to Figure-12.

Figure-12. View selected for projection

Auxiliary View

The **Auxiliary View** tool is used to create projected view from a view by making selected edge of view parallel to screen. The procedure to use this tool is given next.

• Click on the **Auxiliary View** button from the **Ribbon**. The **Auxiliary View PropertyManager** will display and you will be asked to select an axis, edge or sketch line to generate the view.

• Select the axis/edge/sketch line. The preview of view will be displayed; refer to Figure-13.

Figure-13. Auxiliary view

Section View

The **Section View** tool is used to create section view by cutting the model using section lines. The procedure to use this tool is given next.

• Click on the **Section View** button from the **Ribbon**. The **Section View Assist PropertyManager** will display; refer to Figure-14.

• Select the desired button from the **Cutting Line** area of the **PropertyManager**.

- Click in the view to specify the start point of line. The toolbar will display; refer to Figure-15.

Figure-15. Section view toolbar

Figure-14. Section View Assist
PropertyManager

- Select the desired button from toolbar to change the style of cutting line and specify the points.
- Click on the **OK** button from the toolbar, the preview will display; refer to Figure-16.
- Click to place the view. The **Section View PropertyManager** will display; refer to Figure-17.

Figure-16. Preview of section view

Figure-17. Section View PropertyManager

- Set the desired option in the **PropertyManager**. Note that from SolidWorks 2017 onwards, you can select the **Emphasize outline** check box in the **Section View** rollout of **PropertyManager** to create sections compliant to ISO 128-50.
- Click on **OK** button from the **PropertyManager** to create the section view.

Detail View

The **Detail View** tool is used to create detailed view of a specific portion of the model. The steps to create detail view are given next.

- Click on the **Detail View** button from the **Ribbon**. The **Detail View PropertyManager** will display. Also, you are asked to specify center point of circle for creating detail view circle.
- Click in the drawing view at desired location to specify center of the detail circle.
- Click to specify the radius of the circle. Preview of detail view will be attached to the cursor.
- Change the scale of the detail view by using the options in the **PropertyManager**; refer to Figure-18.
- Select the desired style of detail view circle from the **Style** drop-down in the **Detail Circle** rollout of **PropertyManager**; refer to Figure-19.

Figure-19. Styles of detail circle

Figure-18. Setting scale of detail view

- Select the desired check box from the **Detail View** rollout of the **PropertyManager** to change the outline of detail view.
- Click to place the detail view. Press **ESC** to exit the tool.

Broken-out Section

The **Broken-out Section** tool is used to create section of the selected view to display inner detail of the model. Note that you can use the tool on section views and detail views as well. Follow the steps to use this tool.

- Click on the **Broken-out Section** tool from the **Ribbon**. You are asked to specify the start point of the section spline.
- Click in the view to start spline and create a close spline; refer to Figure-20.

Figure-20. Spline for broken-out section

- Click on the **Preview** check box and specify the depth in the spinner in the **PropertyManager**. The preview of broken-out section will display; refer to Figure-21.

Figure-21. Preview of broken-out section

- Click on the **OK** button to create the section.

Break

The **Break** tool is used to represent very long objects in the drawing by breaking them at specific span. The procedure to use this tool is given next.

- Click on **Break** tool from the **Ribbon**.
- Click on the view that you want to break. The break line will attach to the cursor; refer to Figure-22.
- Set the desired gap and break line style in the **PropertyManager**, and click to specify starting point of break span.
- Click to specify the end of break span. The selected span will be removed and the broken view will display with specified gap; refer to Figure-23.

Figure-23. Broken View

Figure-22. Breakline attached to cursor

- Click on the **OK** button to create the view.

ADDING ANNOTATIONS TO VIEW

Till this point, we have learned to place views but without dimensions and annotations these views are of no use for manufacturers. So, we will now add annotations to views. The tools to apply annotations are available in the **Annotation** tab of the **Ribbon**; refer to Figure-24.

Figure-24. Annotation tab of Ribbon

Smart Dimension

The **Smart Dimension** tool is used to dimension each entity of the model on paper. This tool and most of the other tools in the **Smart Dimension** drop-down work in the same way as they do in Part environment.

Chamfer Dimension

The **Chamfer Dimension** tool is used to dimension chamfers in the drawing views. The procedure to use this tool is given next.

- Click on the **Chamfer Dimension** tool from the **Smart Dimension** drop-down in the **Annotation** tab of the **Ribbon**. You will be asked to select the edge to be chamfered.
- Click on the chamfered edge. You will be asked to select the reference edge.
- Click on the adjoining reference edge. The dimension will get attached to cursor; refer to Figure-25.

Figure-25. Chamfer dimension attached to cursor

- Click at the desired location to place the chamfer dimension. The **Dimension PropertyManager** will be displayed. The options in this **PropertyManager** are discussed later in this chapter. For the time being press **ESC** to exit the **PropertyManager**.

Model Items

The **Model Items** tool is used to import all the dimensions/annotations applied to the model in Part modeling environment. To import the dimensions/annotations, follow the steps given next.

- Click on the **Model Items** tool from the **Ribbon**. The **Model Items PropertyManager** will display; refer to Figure-26.
- Select the desired buttons from the **PropertyManager** to import the respective annotations in the view.
- Click in the **Source** drop-down in the **Source/Destination** rollout to specify the entities that you want to annotate in the view.
- Click on the **OK** button to generate the annotations from the model. The automatic annotations will be generated. Drag the annotations to place them properly.

Select one of the dimension, the **Dimension PropertyManager** will be displayed. The options in this **PropertyManager** are same as discussed in **Advanced Dimensioning Chapter**.

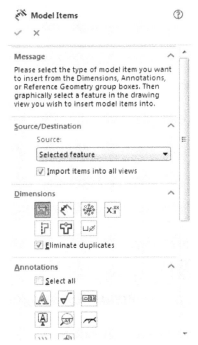

Figure-26. Model Items PropertyManager

Note

The **Note** tool is used to specify extra information in the drawing that are not mentioned in the dimensions. For example, if you want to say "All dimensions are in mm" then this is the tool to do so. The steps to use this tool are given next.

- Click on the **Note** tool from the **Ribbon**. The **Note PropertyManager** will display and note box will get attached to the cursor; refer to Figure-27.

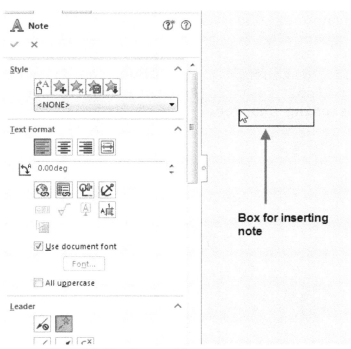

Box for inserting note

Figure-27. Note PropertyManager

- If you move the cursor over any entity in the view, then the leader will be added before the note box; refer to Figure-28.

Figure-28. Note box with leader

- Click to place the box. If leader is attached then click again to place the note box.
- On clicking, the editing mode of note will activate. Enter the desired text in the box. Apply desired formatting by using the options displayed.
- Click on the **OK** button to create the note.

Flag Notes

The Flag Notes are used to cross-reference listed notes to specific area or feature in the drawing. In other words, flag notes are numbered list of items referenced to different area of drawing. The procedure to create flag notes is given next.

- Click on the **Note** tool from the **Ribbon** and click in the drawing area to specify insertion point.
- Click on the **Number** button from the toolbar displayed; refer to Figure-29.

Figure-29. Number button in toolbar

- Type the desired note in the text box in the form of numbered list; refer to Figure-30.

Figure-30. List of notes created

- Click on the number in the list which is to be added in the Flag Note Bank and select the **Add to Flag Note Bank** check box from the **PropertyManager** displayed; refer to Figure-31.

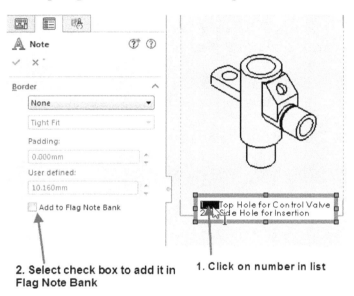

Figure-31. Adding notes flag note bank

- Similarly, you can add other numbered notes in the Flag Note Bank.
- Click on the **OK** button from the **Note PropertyManager** after adding all the notes in the Flag Note Bank.

Now, we will add balloons in the drawing as per the notes. The procedure to add balloons is given next.

- Click on the **Balloon** tool from the **Annotation** tab in the **Ribbon**. The **Balloon PropertyManager** will be displayed; refer to Figure-32.

Figure-32. Balloon PropertyManager

- Select the **Flag Note Bank** check box. The notes earlier saved will be displayed in the list; refer to Figure-33.

Figure-33. Flag Note Bank

- Select the desired note from the list. A balloon will get attached to the cursor.
- Click on the desired reference in the drawing and place the balloon; refer to Figure-34.

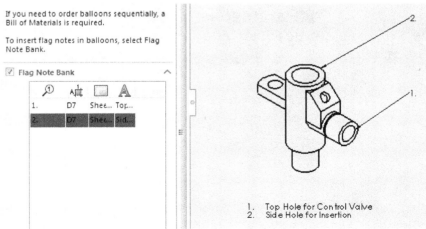

Figure-34. Flag notes created

- Click on the **OK** button from the **Balloon PropertyManager** to exit.

Note that using the options in the **Note PropertyManager**, you can generate all type of geometric annotations like, Datum reference, surface finish symbol, Geometric Tolerance box and so on. The options in the **Note PropertyManager** are discussed next.

Style Rollout

You can save a note as favorite using the options in the **Style** rollout. The options available in this rollout are the same as discussed earlier.

Text Format Rollout

The **Text Format** rollout is used to set the format of the text such as font, size, justification, and rotation of the text. You can also add symbols and hyperlinks to the text using the options available in this rollout.

Leader Rollout

The options in the **Leader** rollout are used to define the style of arrows and leaders that are displayed in the notes.

Leader Style Rollout

The options in this rollout are used to define the style and thickness of the leader. By default, the **Use document display** check box is selected. So, the leader will be displayed with the default style and thickness. On clearing this check box, the **Leader Style** and **Leader Thickness** drop-down lists will be enabled. Using these drop-down lists, you can specify different styles and thickness for the leader.

Border Rollout

The options in the **Border** rollout are used to define the border in which the note text will be displayed. You can assign various types of borders from the **Style** drop-down list. The **Size** drop-down list available in this rollout is used to define the size of the border in which the text will be placed.

Parameters Rollout

The **Parameters** rollout is used to specify the X and Y coordinate values of the note center.

Wordwrap Rollout

Select the **Wordwrap** check box to expand the **Wordwrap** rollout. Specify the desired wordwrap width in the edit box.

Layer Rollout

This rollout is used to assign existing layer or create new layer to the notes.

Surface Finish

The **Surface Finish** tool is used to add surface finish symbol to the surface of a part. The procedure to use this tool is given next.

- Click on the **Surface Finish** tool from the **Annotation CommandManager** in the **Ribbon**. The surface finish symbol will get attached to cursor and the **Surface Finish PropertyManager** will be displayed; refer to Figure-35.

Figure-35. Surface Finish PropertyManager

- Select the desired symbol button from the **Symbol** rollout of the **PropertyManager**.
- Specify the respective parameters in the **Symbol Layout** rollout.
- Set the orientation of symbol from the **Angle** rollout.
- Click at the desired location in drawing view to place the symbol.
- Click on the **OK** button from the **PropertyManager** to exit the tool.

Surface Texture Symbols or Surface Roughness Symbols

Datum Feature

The **Datum Feature** tool is used to add reference for measuring all the geometric tolerances. For example, if you want to check perpendicularity of a face then you need to give a reference with respect to which the perpendicularity will be measured. The steps to create datum features are given next.

- Click on the **Datum Feature** button from the **Ribbon**. The datum symbol will attach to the cursor and the **Datum Features PropertyManager** will display; refer to Figure-36.

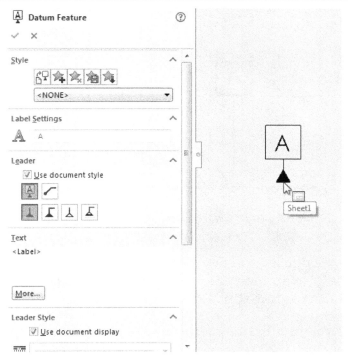

Figure-36. Datum Feature PropertyManager

- Click on the reference to place the leader's start point. Move the cursor to specify length of the leader.
- Click to place the label of the datum feature. Next, datum feature box will attach to the cursor. Press **ESC** to exit the tool. Refer to Figure-37 for datum feature placement.

Figure-37. Datum feature placement

Figure-38 shows the break-up of a datum feature symbol.

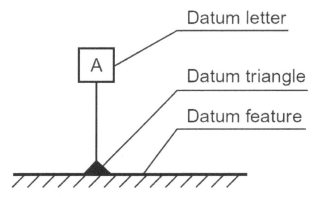

Figure-38. Datum feature symbol

Datum Target

Datum targets are circular frames divided in two parts by a horizontal line. The lower half represents the datum feature, and the upper half is for additional information, such as dimensions of the datum target area; refer to Figure-39.

Figure-39. Datum target symbol

The **Datum Target** tool is used to add datum target to the drawing. The steps to do so are given next.

• Click on the **Datum Target** button from the **Ribbon**. The **Datum Target PropertyManager** will display; refer to Figure-40.

Figure-40. Datum Target PropertyManager

- Specify the desired value and click in the drawing view to place the leader start point.
- Stretch the symbol and click to place at the desired position. Note that you can now make the move or rotate the symbol while placing by using the **Horizontal Movable** ⊖ button or **Rotational Movable** ⊖ button.

Geometric Tolerance

The feature control frame is also known as GD&T box in laymen's language. The method to insert Feature Control Frame in drawing is same as discussed for Surface Finish symbol. In GD&T, a feature control frame is required to describe the conditions and tolerances of a geometric control on a part's feature. The feature control frame consists of four pieces of information:

1. GD&T symbol or control symbol
2. Tolerance zone type and dimensions
3. Tolerance zone modifiers: features of size, projections…
4. Datum references (if required by the GD&T symbol)

This information provides everything you need to know about geometry of part like what geometrical tolerance needs to be on the part and how to measure or determine if the part is in specification; refer to Figure-41. The common elements of feature control frame are discussed next.

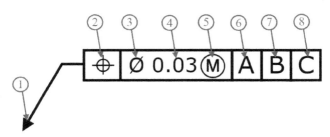

Figure-41. Feature control frame

1. **Leader Arrow** – This arrow points to the feature that the geometric control is placed on. If the arrow points to a surface than the surface is controlled by the GD&T. If it points to a diametric dimension, then the axis is controlled by GD&T. The arrow is optional but helps clarify the feature being controlled.
2. **Geometric Symbol** – This is where your geometric control is specified.
3. **Diameter Symbol (if required)** – If the geometric control is a diametrical tolerance then the diameter symbol (Ø) will be in front of the tolerance value.
4. **Tolerance Value** – If the tolerance is a diameter you will see the Ø symbol next to the dimension signifying a diametric tolerance zone. The tolerance of the GD&T is in same unit of measure that the drawing is written in.
5. **Feature of Size or Tolerance Modifiers (if required)** – This is where you call out max material condition or a projected tolerance in the feature control frame.
6. **Primary Datum (if required)** – If a datum is required, this is the main

datum used for the GD&T control. The letter corresponds to a feature somewhere on the part which will be marked with the same letter. This is the datum that must be constrained first when measuring the part. Note: The order of the datum is important for measurement of the part. The primary datum is usually held in three places to fix 3 degrees of freedom

7. **Secondary Datum (if required)** - If a secondary datum is required, it will be to the right of the primary datum. This letter corresponds to a feature somewhere on the part which will be marked with the same letter. During measurement, this is the datum fixated after the primary datum.

8. **Tertiary Datum (if required)** - If a third datum is required, it will be to the right of the secondary datum. This letter corresponds to a feature somewhere on the part which will be marked with the same letter. During measurement, this is the datum fixated last.

Reading Feature Control Frame

The feature control frame forms a kind of sentence when you read it. Below is how you would read the frame in order to describe the feature.

Meaning of various geometric symbols are given in Figure-42.

SYMBOL	CHARACTERISTICS	CATEGORY
—	Straightness	Form
▱	Flatness	
○	Circulatity	
⌀	Cylindricity	
⌒	Profile of a Line	Profile
⌓	Profile of Surface	
∠	Angularity	Orientation
⊥	Perpendicularity	
//	Parallelism	
⊕	Position	Location
◎	Concentricity	
⚌	Symmetry	
↗	Circular Runout	Runout
↗↗	Total Runout	

Figure-42. Geometric Symbols

Figure-43 and Figure-44 shows the use of geometric tolerances in real-world.

Figure-43. Use of geometric tolerance 1

Figure-44. Use of geometric tolerance 2

Note that in applying most of the Geometrical tolerances, you need to define a datum plane like in Perpendicularity, Parallelism and so on. There are a few dimensioning symbols also used in geometric dimensioning and tolerances, which are given in Figure-45.

Symbol	Meaning
Ⓛ	LMC – Least Material Condition
Ⓜ	MMC – Maximum Material Condition
Ⓣ	Tangent Plane
Ⓟ	Projected Tolerance Zone
Ⓕ	Free State
⌀	Diameter
R	Radius
SR	Spherical Radius
S⌀	Spherical Diameter
CR	Controlled Radius
⑤Ⓣ	Statistical Tolerance
77	Basic Dimension
(77)	Reference Dimension
5X	Places

Symbol	Meaning
←⊕	Dimension Origin
⎵	Counterbore
∨	Countersink
⤓	Depth
⌀	All Around
↔	Between
✕	Target Point
▷	Conical Taper
◁	Slope
☐	Square

Figure-45. Dimensioning symbols

The steps to create geometric tolerance box are given next.

- Click on the **Geometric Tolerance** button from the **Annotations** tab in the **Ribbon**. The **Geometric Tolerance PropertyManager** will display along with the **Properties** dialog box; refer to Figure-46.
- Click on the **Symbol** drop-down and select the desired button from the box displayed; refer to Figure-47.

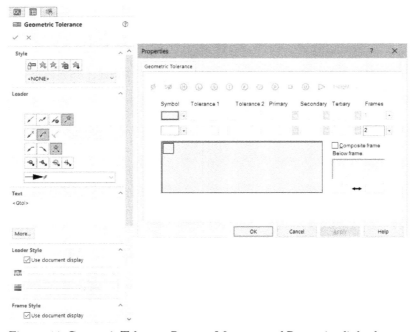

Figure-46. Geometric Tolerance PropertyManager and Properties dialog box

Figure-47. Symbol drop-down

- Click in the **Tolerance** box and specify the deviation.
- Click in the **Primary** box and specify the datum feature label with respect to which the tolerance will be calculated.
- You can specify second tolerance by selecting the **Tolerance 2** check box.
- Click on the **Help** button from the **Properties** dialog box to know more about the box.

Weld Symbol and Hole Callout

The weld symbols are placed in the same way as you place geometric tolerances. Click on the tool, define the parameters and place the symbol by clicking.

To add the hole callout, click on the **Hole Callout** tool and select hole or slot. The callout will be generated automatically. Click to place the callout.

In the same way, you can annotate center line and center mark by selecting the **Centerline** and **Center Mark** tool.

Now, we will add exploded view of the assembly and then we will add bill of material and balloons.

GENERATING EXPLODED VIEW OF ASSEMBLY

To generate exploded view, you must explode the assembly first in the assembly environment. After exploding save the assembly and then follow the steps to generate exploded view.

- Start a new drawing and click on the **Browse** button from the **Model View PropertyManager**.
- Double click on the assembly file that you saved earlier.
- Click on the **Show in exploded state** check box and select the exploded view from the drop-down. Note that to use exploded state view, you must have saved a configuration of assembly in exploded state. The topic has been discuss in previous chapters.
- Select the **Isometric** button from the **Orientation** rollout and click in the drawing area to place the assembly; refer to Figure-48.

Figure-48. Exploded view of assembly

- You can change the scale as per requirement by using the open in the **PropertyManager**.
- Click on the **OK** button to create the view.

Generating Bill of Material

- Click on the **Annotation** tab and click on the down arrow of **Tables** at the right corner; refer to Figure-49.

Figure-49. Tables drop-down

- Click on the **Bill of Materials** button. You are asked to select the view.
- Select the exploded view. The **Bill of Materials PropertyManager** will display.
- Click on the **OK** button from the **PropertyManager**. The Bill of Materials table will attach to the cursor.
- Click to place the table; refer to Figure-50.

Note that in the same way, you can use the other table options in the **Tables** drop-down.

ITEM NO.	PART NUMBER	DESCRIPTION	WEIGHT	QTY.
1	Body			1
2	Washer			1
3	Valve spindle			1
4	Spring			1
5	Spring cap nut			1
6	Pre adjusting nut			1
7	Nozzle valve			1
8	Nozzle body			1
9	Cap nut			1

Figure-50. Bill of Materials

Generating Balloons for Bill of Material

- Click on the **Auto Balloon** button from the **Ribbon** and click on the view. The balloons will be generated automatically.
- Click on the **OK** button from the **PropertyManager** to generate the balloons.
- Drag the balloons to desired positions.

You can use the **Sketch** tab and use the sketcher tools to create custom entities in the drawing.

EDITING TITLE BLOCK

You can edit the title block to specify the information related to designer, part name, versions and so on. To edit the title block, follow the steps given next.

- Right-click on the **Sheet Format** from the **FeatureManager Design Tree**. The shortcut menu will display; refer to Figure-51.

Figure-51. Right click shortcut menu

- Click on the **Edit Sheet Format** button, the title block will become edit able; refer to Figure-52.
- Double-click on the empty boxes in which you want to specify informations.

Figure-52. Editable title block

- After specifying the information, click on the **OK** button from the **PropertyManager**.
- Similarly, you can edit the existing information by double-clicking on them.
- Click on the **Return** button at the top-right corner of the graphic window to return in Drawing mode; refer to Figure-53.

Figure-53. Return to drawing

Practical

In this practical, you will first create the model of part as per the production drawing given in Figure-54 and then you will create the same production drawing of part using the model.

Figure-54. Production drawing for practical

Start a New Part

- Start SolidWorks if not started yet. Press CTRL+N from keyboard.
- Double-click on the **Part** button from dialog box. The Part environment will be displayed.

Creating Part

From the drawing we can see that the whole part is a revolve feature with holes created on it.

- Click on the **Revolved Base/Boss** tool from the **Features CommandManager**. You will be asked to select the sketching plane.
- Click on the **Front Plane** to create sketch on front plane.
- Draw the sketch as shown in Figure-55 while taking the base dimensions from the drawing.
- Select the **75** dimension from the sketch. The **Dimension PropertyManager** will be displayed.
- Click on the **Diameter** button from the **Dimension Text** rollout of the **PropertyManager**. A diameter symbol will be added to the dimension text; refer to Figure-56.

Figure-55. Sketch for revolve feature

Figure-56. Diameter symbol added to dimension text

- Select the **Symmetric** option from the **Tolerance Type** drop-down in the **Tolerance/Precision** rollout of **PropertyManager** and specify the value of tolerance as **0.50** in the respective edit box. The dimension will be displayed as shown in Figure-57.
- Click on the **OK** button from the **PropertyManager**.
- Select the other dimensions one by one and repeat the procedure to specify diameter symbol and tolerances as required. After applying symbols and tolerances the sketch should be displayed as shown in Figure-58.

Figure-57. Tolerance applied to dimension

Figure-58. Sketch after applying tolerances and symbols

- Click on the **Exit Sketch** button from the **CommandManager**. You will be asked to select the axis of revolution.
- Select the centerline created in sketch. Make sure the revolution angle is set to **360** degree.
- Click on the **OK** button from the **Revolve PropertyManager**. The revolve feature will be created; refer to Figure-59.

Figure-59. Revolve feature created

- Click on the Hole Wizard tool and create a drill hole of diameter **6** at the top quadrant point of a construction circle Ø40 on the face revolve feature; refer to Figure-60.

Figure-60. Drill hole created

- Create the circular pattern of hole equidistant on PCD with 3 instances; refer to Figure-61.

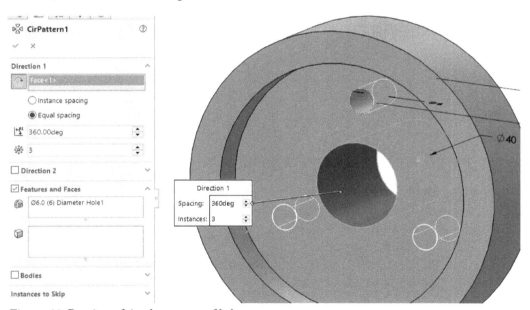

Figure-61. Preview of circular pattern of hole

- Click on the **Thread** tool from the **Hole Wizard** drop-down in the **Features CommandManager** and create the **Metric Die** thread on **M30x2.5**; refer to Figure-62.
- Save the part with the name **Drawing Practical**.

Figure-62. Preview of thread on revolve feature

Starting Drawing File

- Press **CTRL+N** from keyboard. Double-click on the **Drawing** button from the **New SOLIDWORKS Document** dialog box displayed. The Drawing environment will be displayed along with **Sheet Format/Size** dialog box.

- Select the **A4 (ANSI) Landscape** option the list box and click on the **OK** button. (Remember that in First Chapter of the book, we have selected **ANSI** as default drafting standard so we are displayed only ANSI templates of sheet. Click on the **Browse** button from the dialog box to access other templates.)

- On clicking the **OK** button from the **Sheet Format/Size** dialog box; the **Model View PropertyManager** is displayed with open part selected in the box.

Inserting the Base View

- Click on the **Next** button ⊕ from the **PropertyManager**. The front view of model will get attached to cursor.

- Select the **Left** button from the **Orientation** rollout of the **PropertyManager** and place the view near left edge of the drawing sheet; refer to Figure-63.

Figure-63. Placing left view

*   Press **ESC** to exit the tool.

Creating Section View

*   Click on the **Section View** tool from the **View Layout CommandManager** in the **Ribbon**. The section line will get attached to cursor.
*   Click in the center of the view. A toolbar will be displayed; refer to Figure-64.

Figure-64. Specifying section line position

*   Right-click to accept the section line. The section view will get attached to cursor.
*   Place the view at a suitable distance from base view so that all annotations to be applied later can accommodate; refer to Figure-65. Press **ESC** to exit the tool.

Figure-65. Placing section view

Applying Annotations

*   Click on the **Model Items** tool from the **Annotation CommandManager**. The **Model Items PropertyManager** will be displayed.
*   Set the options in the **PropertyManager** as shown in Figure-66.

*Figure-66. Options selected for model
item annotations*

- Click on the **OK** button from the **PropertyManager**. The dimensions will be placed in the drawing views.
- Drag the dimensions to suitable places. Delete extra dimensions and use the **Smart Dimension** tool to create dimensions which are left; refer to Figure-67.

Figure-67. Drawing after editing dimensions

- Add the **Geometric Tolerance** and **Datum Feature** symbols in the drawing as per Figure-54.

PRACTICE 1

Generate the drawing views of all the solid models we have created in Practices and Practicals in the previous chapters.

PRACTICE 2

Generate the exploded views, bill of materials and balloons from all the assembly models we have worked on till this chapter.

PRACTICE 3

Create the model and drawing given in Figure-68.

Figure-68. Practice 3

PRACTICE 4

Create the model and drawing given in Figure-69.

Figure-69. Practice 4

Note that the drawing given in this book are for practice purpose only.

SELF ASSESSMENT

Q1. In which dialog box, we can set the projection type of views being inserted in drawing?

a. **Sheet Format/Size** dialog box
b. **Sheet Properties** dialog box
c. **New SolidWorks Document** dialog box
d. **Geometric Tolerance** dialog box

Q2. In which of the following tab, the tools to add views in the drawing are available?

a. **View Layout** tab
b. **Annotation** tab
c. **Evaluate** tab
d. **Sheet Format** tab

Q3. We can insert as many model views of a part as required. (T/F)

Q4. We can insert the models views of different parts in same drawing by using the **Model View** tool. (T/F)

Q5. The **Auxiliary View** tool is used to create projected view from selected view by making selected edge of view parallel to screen. (T/F)

Q6. The..........tool is used to create section in the selected view to display inner detail of the model.

Q7. The.........tool is used to represent very long objects in the drawing by breaking them at specific span.

Q8. ....is defined as the maximum permissible overall variation of form or position of a feature.

For Student Notes

Chapter 11

AnalysisXpress

Topics Covered

The major topics covered in this chapter are:

- *Perform Simulation Xpress Analysis.*
- *Perform Flow Xpress Analysis.*
- *Perform DFM Xpress Analysis.*
- *Perform Costing of manufacturing process*
- *Perform Sustainability analysis of part*

INTRODUCTION

Analysis Xpress is the combination of tools available in SolidWorks to perform some very useful analyses at a very high pace. There are various types of analyses available in SolidWorks like; SimulationXpress Analysis, FlowXpress Analysis and so on. These analyses are performed for the following functions:

SimulationXpress Analysis: This analysis is used to check whether the component will fail on the specified force/pressure conditions or not. You can generate a report and perform the optimization.

FloXpress Analysis: This analysis is used to check the flow of a fluid through the designed passage.

DFMXpress Analysis: This analysis is used to check whether the created component is manufacturable or not.

StainabilityXpress: This analysis is used to check the impact of environment on the component.

Part Reviewer: This analysis is used to check how the part was created in SolidWorks. This analysis becomes very useful to find error in modeling in SolidWorks.

Along with the above analysis tools, there is one more tool named DriveWorksXpress. This tool is used to increase the speed of designing by making the products formulae based. For example, you can create a formula for Nuts or Bolts and then specify the driving dimensions to create multiple instances of Nuts/Bolts with different sizes.

Note that you need to enter the product codes before using the features discussed in this chapter.

The tools used to perform the named above analyses are discussed next.

SIMULATIONXPRESS ANALYSIS WIZARD

As discussed earlier, the **SimulationXpress Analysis Wizard** tool is used to perform a quick simulation analysis on the model. You can use this tool to perform linear static analysis only. This tool is available in the **Evaluate** tab of the **Ribbon**; refer to Figure-1. The procedure to use this tool is given next.

Figure-1. SimulationXpress_Tool.png

- Click on the **SimulationXpress Analysis Wizard** tool to start SimualtionXpress Analysis, SolidWorks Simulation Xpress pane will be displayed on the right in the program window; refer to Figure-2. **Make sure that you have opened the solid model for which you want to perform the analysis.**
- Click on the **Options** button in the right of the application window; refer to Figure-3. On doing so, the **SimulationXpress Options** dialog box will be displayed as shown in Figure-4.

Figure-2. Interface

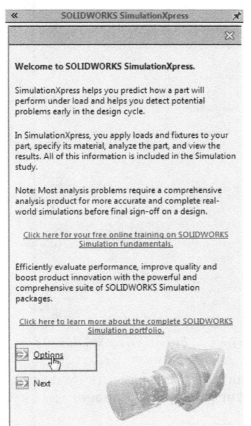

Figure-3. Options button

Figure-4. SimulationXpress Options dialog box

The options in this dialog box are used to set the unit system of the analysis and save directory for the analysis report.

- Select the **SI** option from the **System of units** drop-down and change the location of result report to the desired one by using the Browse button. ⬚ After setting all the parameters, select the **OK** button from the dialog box.

Fixture Setting

- Select the **Next** button from the **SolidWorks SimulationXpress** task pane, the **Fixtures page** of SimulationXpress will be displayed as shown in Figure-5.

Figure-5. Fixtures page of SimulationXpress

Click on the **Add a fixture** button to fix a face. On clicking the **Add a fixture** button, the **Fixture PropertyManager** will be displayed; refer to Figure-6.

Figure-6. Fixture PropertyManager

- Select the face of the model that you want to be fixed; refer to Figure-7.

Face selected

Figure-7. Fixed face

- Click on the **OK** button from the **Fixture PropertyManager** and then click on the **Next** link button from the **SimulationXpress** displayed in the right. On doing so, the **Loads** page will be displayed in the right; refer to Figure-8.

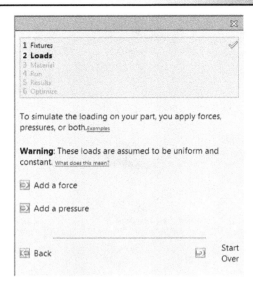

Figure-8. Loads page

Load Setting

- Click on **Add a force** or **Add a pressure** link button from the **SimulationXpress**. In our case, the **Add a pressure** link button is selected. On doing so, the **Pressure PropertyManager** will be displayed as shown in Figure-9.

Figure-9. Pressure PropertyManager

- Select the face of the model on which you want to apply the load; refer to Figure-10.

Figure-10. Face to be selected

- Change the unit from N/mm^2 to psi by clicking on the **Unit** drop-down and selecting the **psi** option, if required.
- Specify the desired pressure value in the **Pressure Value** edit box. After specifying the value, click on the **OK** button from the **PropertyManager**.

You can change the direction of pressure by selecting the Reverse direction check box from the FeatureManager.

- You can add more loads by using the **Add a force** or **Add a pressure** link button. After specifying all the desired loads, click on the **Next** link button from the **SimulationXpress**. On doing so, the **Material** page will be displayed; refer to Figure-11.

Figure-11. Material page of SimulationXpress

Material Setting

- Check the Warning message in the task pane carefully (This is the conditions of using this analysis) and then click on the **Choose Material** link button from the **SimulationXpress**, the **Material** dialog box will be displayed as shown in Figure-12.

Figure-12. Material dialog box

- Note that some of the properties of the material are highlighted in red color. These highlighted properties are the driving properties for the analysis. Select the desired material from the list in the left and then select the **Apply** button from the dialog box.
- Click on the **Close** button to exit the dialog box. The material will be applied and its properties will be displayed in the **SimulationXpress task pane**.

- Select the **Next** button from the **SimulationXpress** Task Pane. On selecting the button, the **Run** page will be displayed; refer to Figure-13.

Changing Mesh Density

- Click on the **Change settings** link button from the **Run** page and then click on the **Change mesh density** link button from the page. The **Mesh PropertyManager** will be displayed; refer to Figure-14.

Figure-13. Run page of SimulationXpress

Figure-14. Mesh Property-Manager

- Click on the **Mesh Parameters** check box in the **PropertyManager** and set the desired value of Global mesh element size as well as the tolerance in the respective edit boxes.

Or

- Drag the **Mesh Density** slider to desired location to make the mesh coarse or fine. Note that the finer you make the mesh, the longer system will take to solve simulation.
- Click on the **OK** button from the **PropertyManager** to apply the changes.
- Click on the **Next** button from the Task Pane to return to Run page.

Running Simulation

- Click on the **Run Simulation** link button from the **SimulationXpress** to check the output. The result will be displayed in the Modeling area and Results page will be displayed as shown in Figure-15.

Figure-15. Results page of Simula-tionXpress

Results

- Select the **Play animation** and **Stop animation** button to start and stop the simulation. If you agree with the result you need to select the **Yes, continue** link button and if you disagree then select the **No, return to Loads/Fixtures** link button to change the parameters of analysis.

- On selecting the **Yes,continue** button, the modified **Results** page will be displayed as shown in Figure-16.

- You can display von Mises stress or displacement by using the respective button from the SimulationXpress Task Pane. Also, the suggested FOS will also be displayed in the SimualtionXpress.

- After checking the results, select the **Done viewing results** link button; the modified results page will be displayed as shown in Figure-17. Now, you can generate and eDrawing or you can generate a report by selecting the respective button from the SimualtionXpress.

Figure-16. Modified results page of
SimulationXpress

Figure-17. Modified Results page

- Select the **Generate report link** button to create the report. On selecting this button, the **Report Settings** dialog box will be displayed as shown in Figure-18.

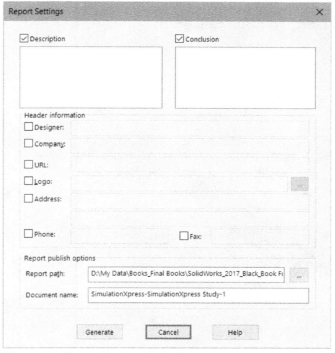

Figure-18. Report Settings dialog box

- Select the check boxes to enter information in the desired field in the dialog box. You can enter the path for saving report in the **Report path** field of the dialog box. After specifying the information select the **Generate** button to create the report at the specified path, the **Generating Report** dialog box will be displayed as shown in Figure-19. Once the report generation gets completed; a word document will open automatically with the complete report of analysis.

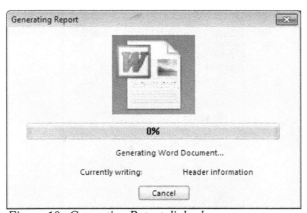

Figure-19. Generating Report dialog box

- Similarly, you can generate an eDrawing of the results report by selecting **Generate eDrawings file** link button. The eDrawings can be opened in the eDrawing software provided with the SolidWorks package; refer to Figure-20.

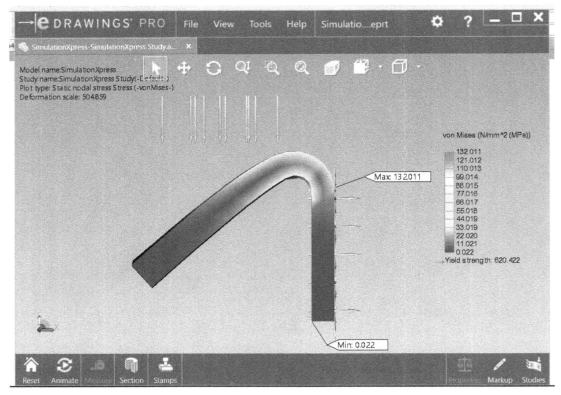

Figure-20. SimulationXpress report in eDrawing

Optimizing

- After checking the results, if you want to optimize the model then select the **Next** link button from the SimualtionXpress, the **Optimize** page of **SimulationXpress** Task Pane will be displayed as shown in Figure-21.

Figure-21. Optimize page of SimualtionXpress

- Make sure that the **Yes** radio button is selected in the page and then select the **Next** button from the SimulationXpress. On doing so, the **Parameters** dialog box will be displayed and the driving dimensions will be displayed in the modeling area; refer to Figure-22 and Figure-23.

Figure-22. Parameters dialog box

Figure-23. Driving dimensions of the model

- Click on one of the driving dimension from the model that you want to change for optimization (like thickness of part) and then select the **OK** button from the dialog box, the value will be added in the **DesignXpress Study** tab displayed below the modeling area; refer to Figure-24.

Figure-24. DesignXpress Study

- To replace the variable for optimization, click on the **Click here to replace variables** drop-down under the **Variables** node in the **DesignXpress Study**. On doing so, the **Parameters** dialog box will be displayed again. Click in the blank **Name** field and select another dimension from the modeling area to replace.
- After adding the desired variable, select the **OK** button from the dialog box. Now, click on the **Click here to add Constraints** drop-down under the **Constraints** node and select the desired constraint from the list. In this case, we have selected **Factor of Safety** as the constraint; refer to Figure-24.
- Set the value for constrain using the spinner.
- After setting all the desired parameters; click on the **Run** button displayed at the top in the **DesignXpress Study** tab. On doing so, the **DesignXpress Study in Progress** dialog box will be displayed; refer to Figure-25.

Figure-25. DesignXpress Study in Progress dialog box

- After the study is complete, the optimization results will be displayed in the **Results View** tab of the **DesignXpress Study**; refer to Figure-26.

	Initial	Optimal
D1SheetMetal1 (0.00449995)	3mm	4.5mm
Factor of Safety	10.344095	11.587700
Mass	0.415982 kg	0.412314 kg

Figure-26. Results View tab of DesignXpress Study

- In the **Optimize** page of **SimulationXpress**, the options will be modified as shown in Figure-27.

*Figure-27. Modified Optimize
page of SimulationXpress*

- Select the radio button for **Optimal** value and then click on the **Run the optimization** link button again to cross check and then click on the **Next** button to apply the optimization.
- Exit the **SimulationXpress** by clicking on the **Close** button at the top-right corner of it and select **Yes** from the dialog boxes displayed to save results.

FLOXPRESS ANALYSIS

FloXpress Analysis is used to check the flow of a fluid through the designed passage. This passage can be in a solid model or it can be in an assembly. This analysis can be performed by using the **FloXpress Analysis Wizard** button available in the **Evaluate** tab of the **Ribbon**; refer to Figure-28.

Figure-28. FloXpress Analysis Wizard button

To perform the express flow analysis of a passage, you need to close all the openings in the model with the help of a lid. To do so, perform the following steps.

- Open the model for which you want to perform the flow analysis; refer to Figure-29. Note that using the FloXpress you can check flow of component having only one inlet and one outlet.

Figure-29. Model for flow analysis

Preparing Model

- Now, we need to close both the ends. To do so, select the outer round edge of one end of the model and then select the **Filled Surface** tool from the **Surfaces CommandManager** in the **Ribbon**; refer to Figure-30. On doing so, the **Filled Surface PropertyManager** will be displayed. Click on the **OK** button from the **PropertyManager**. Similarly, create the filled surface on the other end. The model after creating the filled surfaces will be displayed as shown in Figure-31.

Figure-30. Filled Surface tool

Model before filled surfaces   Model after creating filled surfaces

Figure-31. Filled surfaces creation

- Now, we need to thicken these surface to create the lid. To do so, select one of the surfaces and then select the **Thicken** tool from the **Surfaces** tab. On doing so, the **Thicken PropertyManager** will be displayed.

- Select the **OK** button from the **PropertyManager**, the surfaces will be thickened. Similarly, thicken the other surface, the model after thickening will be displayed as shown in Figure-32.

Figure-32. Thickened surfaces to create lids

Starting Flow analysis

- Select the **FloXpress Analysis Wizard** tool from the **Evaluate** tab of the **Ribbon**, the **Welcome PropertyManager** will be displayed in the left of the application window; refer to Figure-33.

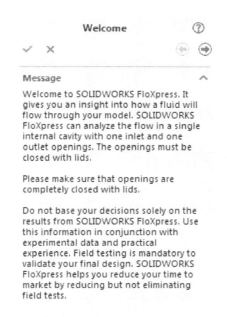

Figure-33. Welcome PropertyManager

- Click on the **Next** button from the **PropertyManager**. On doing so, the **Check Geometry PropertyManager** will be displayed as shown in Figure-34.

Figure-34. Check Geometry PropertyManager

- Click on the **View fluid volume** to check the fluid volume in designed passage. On doing so the model will be displayed as shown in Figure-35.

Figure-35. Fluid Volume in model

- You can change the smallest flow passage for the model at this stage. To do so, click on the **Smallest Flow Passage** button and then change the value by using the spinner.
- After specify the desired parameters, click on the **Next** button from the **PropertyManager**. On doing so, the **Fluid PropertyManager** will be displayed and you are prompted to specify the type of fluid for the analysis. You can select either water or air.
- In this case, we select the **Air** radio button. Now, click on the **Next** button from the **PropertyManager**. On doing so, the **Flow Inlet PropertyManager** will be displayed as shown in Figure-36.

Figure-36. Flow Inlet PropertyManager

- Select the desired button and then specify the desired values in the edit boxes displayed at the bottom of the **Flow Inlet PropertyManager**.
- Select the inner face of the model as inlet; refer to Figure-37. You might need to right-click on the face and then select the **Select Other** option from the shortcut menu.

Figure-37. Face to be selected as Inlet

- After selecting the face, select the **Next** button from the **PropertyManager**. On doing so, the **Flow Outlet PropertyManager** will be displayed as shown in Figure-38.

Figure-38. Flow Outlet PropertyManager

- Now, you need to select the outlet for the model. To do so, move the cursor over the flat face of the outlet lid and then right-click; a shortcut menu will be displayed. Now, select the face as shown in Figure-39.

Figure-39. Face to be selected as Outlet

- To check the flow fluid, you need to change the display style of the model. To do so, select the **Wireframe** button from the **Display Style** drop-down; refer to Figure-40.

Figure-40. Wireframe button

- After changing the display style, the model will be displayed as shown in Figure-41.

Figure-41. Model in wireframe style.jpg

• Click on the **Next** button from the **PropertyManager** to display the **Solve PropertyManager**; refer to Figure-42.

Figure-42. Solve PropertyManager

• Click on the **Solve** button to run the analysis. On doing so, the system will start to solve and after the CFD problem is solved, the solution will be displayed.

• You can generate the report of result by using the **Generate Report** button under the **Report** rollout. On doing so, the word document of report will be generated.

DFMXPRESS ANALYSIS

The DFMXpress analysis is used to check whether the model in the Modeling area is manufacturable or not. Using this analysis, you can check whether the model is manufacturable by Mill/Drill Manufacturing process, Turn with Mill/drill process, Injection Molding process or Sheetmetal process. The procedure is given next.

• To analyze the model for these manufacturing processes, click on the **DFMXpress Analysis Wizard** tool; the **DFMXpress** task pane will be displayed in the right of the application window; refer to Figure-43.

Figure-43. DFMXpress

- Now, click on the **Settings** button to specify the parameters related to the analysis; refer to Figure-44.

Figure-44. Settings page of DFMX-press

- Select the desired process and then specify the parameters in the below fields. These parameters are basically the limitations of your machines to manufacture an item.
- After specifying all the parameters, click on the **Run** button to run the analysis. After the analysis gets completed, the results will be displayed at the bottom in the DFMXpress; refer to Figure-45.

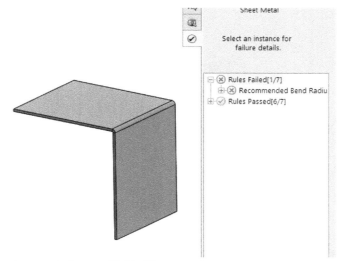

Figure-45. Result of DFMXpress analysis

- Expand the rules which have failed and update your part based on the failed rules.
- Click on the **Close** button from the Task Pane to exit.

COSTING

In real world, it is possible to make same part with two or more processes, like part in Figure-46 can be manufactured by milling, shaping, sheetmetal processes or casting process. But in industry, we are also concerned about the cost of manufacturing and quality of product. In this section, we will learn to estimate the cost of manufacturing the model by various processes.

Figure-46. Model for costing

- Click on the **Costing** button from the **Evaluate CommandManager** in the **Ribbon**. The Costing interface will be displayed; refer to Figure-47.

Figure-47. Costing interface

- Select the desired process from **Method** drop-down; refer to Figure-48.

Figure-48. Method drop-down

- Click in the **Template** drop-down in **Costing** task pane to select the desired template. Using the **Launch Template Editor** button you can launch the **Costing Template Editor** to edit the template; refer to Figure-49. Note that using the Editor you can change the cost and time of various manufacturing process with respect to the selected material. Like, you can change the feed rate for milling a steel material by clicking on the **Mill** tab in right of the **Costing Template Editor** dialog box. Changing the feed rate will change the cycle time and hence the cost of machining.

Figure-49. Costing Template Editor

- After editing the template, save the file and exit the editor.
- Click on the **Begin Cost Estimate** button from the **Message** rollout of the **Costing Task Pane**. Cost estimate will be displayed task pane and breakup of cost will be displayed in the **CostManager**.
- It might be possible that you have some processes which are not having price rate. In that case, you need to manually specify the price. Such processes are displayed under **No Cost Assigned** node; refer to Figure-50.

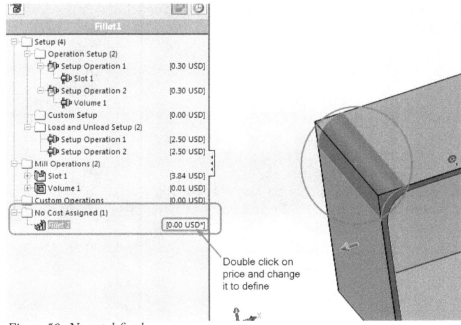

Figure-50. No cost defined

- After specifying all the costs, change the material to the desired one and also update the per kg price of material in the **Material** rollout in task pane; refer to Figure-51.

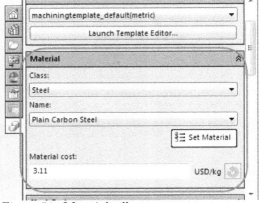

Figure-51. Material rollout

- Check the price of manufacturing in the **Estimated Cost per Part** area of **Costing** task pane and get the optimum process for manufacturing your model.

SUSTAINABILITY

The **Sustainability** tool is used to perform screening-level life cycle assessment (LCA) directly on individual part designs to help you understand the environmental impacts of your design decisions. The procedure to use this tool is given next.

- Click on the **Sustainability** tool from the **Evaluate CommandManager** in the **Ribbon**. The **Sustainability Task Pane** will be displayed in the right of application window; refer to Figure-52.
- Select the desired material from the **Material** rollout of the task pane.
- Scroll down to **Manufacturing** rollout and select the region in which you are manufacturing the part from the Region drop-down; refer to Figure-53.
- Set the other parameters in the **Manufacturing** rollout, **Use** rollout, **Transportation** rollout, and **End of life** rollout of the task pane. The result of analysis will be displayed in the form of four pie charts in the task pane

Figure-52. Sustainablility Task Pane

Figure-53. Manufacturing rollout

- Move the cursor of over the **Current** graph line in the **Carbon** area of the graph. The current value of carbon emission will be displayed; refer to Figure-54.

Figure-54. Current value of carbon emission

- Move the cursor over Carbon pie chart to check the percentage of emission done by various stages of part.
- Similarly, you can check the other environmental impacts of part.
- Change the material and other manufacturing aspects to compare the results.
- Click on the **Close** button on the task pane to exit the tool.

PRACTICE 1

Design a wall bracket to bear a load of 100 kg. There should be two counterbore holes for M10 bolts to mount it on wall. Material for wall bracket is Cast Iron. Estimate the cost of manufacturing wall bracket using

SELF-ASSESSMENT

Q1. Which of the following analysis should be used to test the load bearing capacity of a component?

a. FloXpress Analysis
b. SimulationXpress Analysis
c. StainabilityXpress Analysis
d. DFMXpress Analysis

Q2. Which of the following analysis is used to check the flow of a fluid through the designed passage?

a. FloXpress Analysis
b. SimulationXpress Analysis
c. StainabilityXpress Analysis
d. DFMXpress Analysis

Q3. Which of the following analysis is used to check whether the model in the Modeling area is manufacturable or not?

a. FloXpress Analysis
b. SimulationXpress Analysis
c. StainabilityXpress Analysis
d. DFMXpress Analysis

Q4. FloXpress Analysis can be performed with water as well as air. (T/F)

Q5. Different manufacturing processes hardly affect the production of a component in manufacturing facility. (T/F)

Chapter 12

Mold Tools

Topics Covered

The major topics covered in this chapter are:

- *Starting the Mold tools*
- *Analyzing the model for molding*
- *Preparing the model for mold*
- *Starting the Mold project*
- *Creating the parting line*
- *Creating the Shutoff surfaces*
- *Creating the Parting surface*
- *Creating the Splitting the Core and Cavity from _ _ _ _ _ _g*

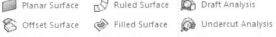

Planar Surface Ruled Surface Draft Analysis Split Line Scale Insert Mold Folders Parting Lines Shut-off Surfaces Parting Surfaces Tooling Split Core

Offset Surface Filled Surface Undercut Analysis Draft

Radiate Surface Knit Surface Parting Line Analysis Move Face

Features Sketch Sheet Metal Mold Tools Evaluate DimXpert SOLIDWORKS Add-Ins SOLIDWORKS MBD

STARTING THE MOLD TOOLS

To start with the mold tools, first you need to open or import a model. To open/ import a model follow steps given next.

- Go to the **File** menu and click on the **Open** option or click on the **Open** button from the **Menu Bar** or click on the **Open a Document** link in the **SolidWorks Resources Task Pane**; refer to Figure-1. The **Open** dialog box will be displayed as Figure-2.

Figure-1. Open or Import methods

Figure-2. Open dialog box

- Click on the **SolidWorks Files** drop-down displayed at the bottom right in the dialog box. List of file types supported by SolidWorks will be displayed; refer to Figure-3.
- Select the type for your file from the list and browse to the folder in which you have placed the file.
- Double-click on the file, you want to open or import. The file will open in the application; refer to Figure-4.

Figure-3. File types
supported

Figure-4. Opened file

- Click on **Mold Tools CommandManager** of **Ribbon** to display tools of molding. If the **Mold Tools CommandManager** is not displaying by default, then right-click on any of the tab displaying in the **Ribbon**. On right-clicking a menu will be displayed; refer to Figure-5.

Figure-5. CommandManager menu

- Click on the **Mold Tools** option from this list, the **Mold Tools CommandManager** will get added in the **Ribbon**. On clicking the **Mold Tools CommandManager**, the Mold Tools will be displayed as shown in Figure-6.

Figure-6. Mold tools tab

- Most of the tools have already been discussed in the book. Now, you will learn about the tools that are specifically used for making mold.

Now, we have the model opened and we need to analyze it for the possibility of its mold design.

ANALYZING THE MODEL

There are three tools available in the **Mold Tools** tab to analyze the model for molding: **Draft Analysis, Undercut Analysis** and **Parting Line Analysis**. These tools are discussed next.

Draft Analysis

The **Draft Analysis** tool is used to check the draft angles of various faces in the model. Draft is an important requirement of Molding. Draft is the taper angle given to various faces of the mold part for easy and safe ejection from the mold tooling (core and cavity). The angle value of draft depends on the material and geometry of the mold part. Typically 1° to 3° of draft is given on all faces of the mold part. If you have steps at parting line, then 5° to 7° of draft is required for shutoff. (**Shutoff** is the surface where core and cavity meet each other.) To perform the draft analysis, follow the steps given next.

- Click on the **Draft Analysis** tool in the tab, **Draft Analysis PropertyManager** will be displayed in the left of the screen; refer to Figure-7.
- A box will be highlighted in blue color in the **Analysis Parameters** rollout of the **PropertyManager**. You are asked to select a face with respect to which the angles will be measured and the pull direction will be defined.
- Select a flat face from the model. Note that this face will become the mating face of core and cavity later. Figure-8 shows the face selected as neutral face.

Figure-7. Draft analysis property-manager

Figure-8. Face to be selected

- As you select the face, all the faces of the model will be painted with the colors specified in the lower area of the **PropertyManager**. By default, the green colored faces are created on core steel and red colored faces are created on cavity steel. The yellow colored faces fall in the undercut categories. These yellow colored faces should either be split up to make them in core & cavity or they should be created on sliders.

- You can change the colors of the faces as per your requirement by selecting the **Edit Color** buttons displayed in the lower area of the **PropertyManager**.
- Note that when you move cursor over the faces of the model, the draft angle value of the current face will be displayed along with the cursor; refer to Figure-9.

Figure-9. Draft angle over the faces

- If you need to adjust the pull direction, then select the Adjustment Triad check box. On doing so, the triad will be displayed around the pull direction arrow; refer to Figure-10. Using this triad, you can change the direction of pull.

Figure-10. Adjustment triad

- Now, check the colors of the model. By default, **Green** color denotes the positive draft which means that the faces are going to be in the **Core**. **Red** color denotes negative draft which means that the faces are going to be in the **Cavity**. **Yellow** color denotes that the faces do not have draft angle applied. **The faces that are painted yellow need to be worked on.** You need to apply the desired draft angle to these faces. The method to apply draft will be discussed later in this chapter.

Undercut Analysis

The **Undercut Analysis** tool is used to check the faces of the mold part that behave as undercuts. SolidWorks classify the undercut faces in five categories: **Direction1 undercut**, **Direction2 undercut**, **Occluded undercut**, **Straddle undercut**, and **No undercut**.

Direction1 undercut: The faces in this category are imprinted on Core tooling.

Direction2 undercut: The faces in this category are imprinted on Cavity tooling.

Occluded undercut: The faces in this category are those faces which can neither be included in core nor in cavity. These faces require sliders. In SolidWorks, sometimes these faces are shifted to No undercut category.

Straddle undercut: The faces in this category are those which you can put in any of the two: the core steel or the cavity steel.

No undercut: The faces in this category are those which are not counted as undercut. But in SolidWorks, sometimes these are the occluded faces. Most of the time, you need to split the face to transfer it in core/cavity steel.

Now, we will perform the Undercut Analysis on the mold part. The steps are given next.

- Exit all the other analyses if still active. Now, click on the **Undercut Analysis** tool. The **Undercut Analysis PropertyManager** will be displayed; refer to Figure-11.
- You can select a pull face or you can select the parting line (if available).
- On selecting the pull face, the mold part will be displayed as shown in Figure-12.

Figure-12. Model with undercut report

Figure-11. Undercut analysis PropertyManager

- You can see from the model and report that there are 10 faces that come in **No undercut** category. If you want to hide all the other faces and want to display only No undercut faces, then click on the **Show/ Hide** button next to all other categories in the **PropertyManager**; refer to Figure-13.

Figure-13. Mold part after hiding other categories

- Sometimes, we need to split these faces to include them in their respective die steel. You will learn about the splitting later in this chapter.

Parting Line Analysis

The **Parting Line** tool is used to check the possible parting line for the mold part. Using this tool, you can check the parting line for multiple pull directions. The steps to perform Parting Line Analysis are given next.

- Check the part given in Figure-14 and try to find out the parting lines manually (Parting line is the line at which the core and cavity meet.)

Figure-14. Part for parting line check

- Now, we will check what should be the parting lines by which we can divide the mold part into core and cavity. To do so, click on **Parting Line Analysis** tool to start the Analysis. The **Parting Line Analysis PropertyManager** will be displayed as shown in Figure-15.

Figure-15. Parting Line Analysis PropertyManager

- Select the faces perpendicular to which the pull direction is to be defined. Refer to Figure-16.

Figure-16. Faces selected for parting line analysis

- Check the dark and light lines. In the above figure, there are three types of lines: dark lines, dashed lines, and dotted lines. Dark lines denote main core and cavity. Dashed lines and dotted lines denote sub-inserts for the mold or the outline where the mold part needs to be divided. (Sub-Inserts are the parts that are assembled in the main Core/Cavity to create a mold part.)

PREPARING MODEL FOR MOLD

After performing the above three analyses, we need to modify the mold part so that molding becomes feasible. There are four options available in SolidWorks **Mold Tools CommandManager** to prepare the model for molding: **Split Line**, **Draft**, **Move Face**, and **Scale**. These tools are discussed next.

Splitting Faces using Split Line tool

In some of the cases, a face of mold part cannot be completely allotted to core or cavity. In those cases, you need to divide that face into two or more parts that can fit for core and cavity. The steps given next explain the use of splitting for mold.

- Perform the **Undercut Analysis** to find out the areas where you need to split the faces to accommodate the face in core/cavity.
- After finding the areas that require splitting, click on the **Split Line** tool from the **Ribbon**. The **Split Line PropertyManager** will be displayed as shown in Figure-17.

Figure-17. Split Line PropertyManager

- By default, **Projection** radio button is selected and you are supposed to select a sketch for dividing the face. Figure-18 shows a mold part that need to be divided from mid.

Figure-18. mold part to be splitted

- A line sketch is drawn at the center of the mold part so that it can divide the part into two pieces; refer to Figure-19.

Note that to draw the sketch, you need to cancel the **Split Line** tool, select the **Sketch** tab from the **Ribbon** and after you draw the sketch, you need to select the **Split Line** tool again.

Figure-19. Sketched line drawn at the center

- Select the line sketch, it will be displayed in the first selection box in the **Split Line PropertyManager**.
- Click in the next selection box. You will be asked to select the faces to be split up.
- Select all the faces of the mold part that you want to split by using the projection of sketch; refer to Figure-20.
- Select the **OK** button from the **PropertyManager**, the mold part will be displayed as shown in Figure-21.

Figure-20. faces selected

Figure-21. Mold part after splitting

- You can also split this part by using the **Silhouette** radio button. Select this radio button, the first selection box will change with **Direction of Pull** selection box.
- Select a plane or face or edge that defines the direction of pull.
- Click in the next selection box and select all the required faces of the mold part that you want to split; refer to Figure-22.

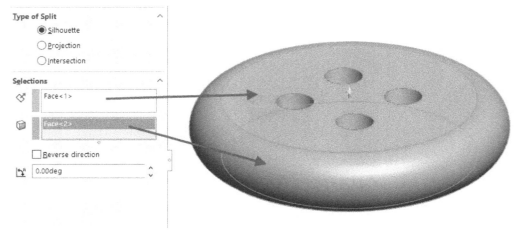

Figure-22. Selection for silhouette split

- Click on the **OK** button, the part will split based on undercut analysis and pull direction; refer to Figure-23.

Figure-23. Silhouette split

- In the same way, you can split the mold part by using the **Intersection** radio button.

Applying draft using Draft tool

After you perform the draft analysis, some of the faces of the mold part will be displayed in yellow color which means that they require the draft. In such cases, we use the **Draft** tool to apply draft angle at those faces. The following steps explain the procedure to apply draft angle.

- Click on the **Draft** tool to apply draft. The **DraftXpert PropertyManager** will be displayed as shown in Figure-24.

Figure-24. DraftXpert PropertyManager

- Select the plane which will act as neutral plane for the draft angle and then select the faces on which you want to apply draft angle; refer Figure-25.

Figure-25. Faces selected for applying draft

- Change the angle value by using the spinner in the **PropertyManager** and then click on the **OK** button to apply the draft.

Increasing/Decreasing thickness of walls
using the Move Face tool

Since it is recommended to have a uniform thickness in the mold part, at some places you may require to change the thickness of part. The steps to change the thickness of wall are given next.

- Click on the **Move Face** tool from the **Ribbon**. The **Move Face PropertyManager** will be displayed as shown in Figure-26.

Figure-26. Move Face PropertyManager

- Select the face that you want to move and then specify the distance value in the **ΔX**, **ΔY**, or **ΔZ** spinners as per your requirement.
- If you want to rotate the faces then click on the **Rotate** radio button and then specify the angle values in the respective spinners at the bottom of the **PropertyManager**.

Figure-27 shows a model while increasing its wall thickness.

Figure-27. Model while increasing wall thickness

Scaling the model to allow shrinkage in part

While creating mold for a part, we need to increase the size of model by certain percentage so that it do not get undersized when it comes out of mold after cooling. (During the cooling of mold part, its plastic shrinks by a certain amount. This certain amount is called shrinkage allowance). The steps to scale a mold part are given next.

- Click on the **Scale** tool, the **Scale PropertyManager** will be displayed as shown in Figure-28.
- By default, **Centroid** is selected in the **Scale about** drop-down. You can select **Origin** or a **Coordinate System** as reference for scaling by using respective option from the **PropertyManager**.
- After selecting the desired option from the drop-down, specify the value of scale in the spinner.

Figure-28. Scale PropertyManager

- Shrinkage value is given by mold designer or material supplier. You need to add the shrinkage value supplied to you in 1. For example, the shrinkage value for ABS plastic is 0.004 supplied to you. In this case, specify the scale factor as 1.004 in the spinner.

After performing the analyses and doing the required operation, we are ready to start the mold project.

INSERTING MOLD FOLDER

This is the first step when you start a mold project in SolidWorks. To start the project, click on the **Insert Mold Folders** button; the folders that are required for various components of mold will be created in the current project.

PARTING LINE

In this step, we design the parting line by using the standard identification of SolidWorks based on draft analysis. Follow the steps given below to create parting line.

- Click on the **Parting Line** tool from the **Ribbon**. The **Parting Line PropertyManager** will display; refer to Figure-29.

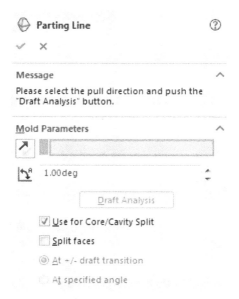

Figure-29. Parting Line PropertyManager

- Click on the neutral plane i.e. a flat face with respect to which the draft angles will be measured.

- Set the draft value in the spinner and click on the **Draft Analysis** button. The model will be colored for core and cavity. Also, the parting line will display in violet color; refer to Figure-30.

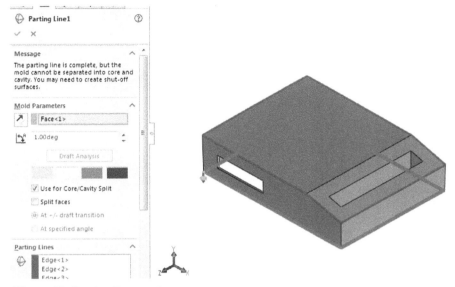

Figure-30. Parting line preview

- Click on the **OK** button from the **Parting Line PropertyManager** to create the parting line.
- If you want to manually create parting line then clear the selection in the **Parting Line**s rollout by using right-click shortcut menu; refer to Figure-31. Now, one by one select the consecutive edges of the model that you want to use as parting line; refer to Figure-32.

Figure-31. Clear Selections option

Figure-32. Edges selected for manual parting line

SHUT-OFF SURFACES

Any opening in the model must be closed by a surface so that the core steel and cavity steel meet at defined surface. The **Shut-off Surfaces** tool performs this job for us. The steps to use this tool are given next.

- Click on the **Shut-off Surfaces** tool from the **Ribbon**. The **Shut-off Surface PropertyManager** will display and the open loop will be selected automatically; refer to Figure-33.

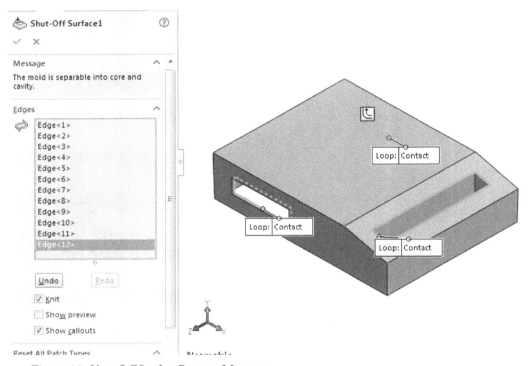

Figure-33. Shut-Off Surface PropertyManager

- Click on the **All Tangent** button from the **Reset All Patch Types** rollout, to create straight patches.
- Click on the **OK** button to create the shut-off surfaces. The surfaces will be created; refer to Figure-32.
- Note that you can also select the edges manually for creating shutoff surfaces in the same way as discussed for parting line.

Figure-34. Shut-off surface to be created for this case

PARTING SURFACES

Parting surface is the surface by which the core and cavity is separated. Parting surfaces are created on the basis of parting lines. Note that parting surface is always a continuous surface. The steps to create parting surface are given next.

- Click on the **Parting Surfaces** tool from the **Ribbon**. The **Parting Surface PropertyManager** will display. Also, the preview of parting surface will display; refer to Figure-35.

Figure-35. Parting surface

- Increase the length of parting surface by using the first spinner in the **Parting Surface** rollout in the **PropertyManager**.
- You can change the direction of parting surface by using the radio buttons in the **Mold Parameters** rollout.
- If the **Perpendicular to pull** radio button is selected then click on the **Manual Mode** check box to manually change the direction of surface by dragging the key points; refer to Figure-36.
- Click on the **OK** button from the **PropertyManager**.

Figure-36. Manual editing of parting surface

TOOLING SPLIT

Now, we want to extract core and cavity from the model by using the parting surface. To split the tooling to generate core and cavity, follow the steps given below.

- Click on the **Tooling Split** tool from the **Ribbon**. You are asked to create sketch for the tooling from which core and cavity will be extracted.
- Select a plane parallel to the parting surface to create tooling; refer to Figure-37.

Figure-37. Face selected for tooling

- On selecting the face, the sketching environment will be displayed.
- Create the sketch for the tooling. You can change the display style to **Hidden Lines Visible** for clear view; refer to Figure-38. Note that the sketch of tooling should be big enough to contain the whole size of component.

Figure-38. Sketch for tooling

- Exit the sketch environment and change the display style to shaded and orientation to isometric; refer to Figure-39.

Figure-39. Extrusion tooling split

- Specify the desired extrusion height in the spinners.
- Click **OK** from the **PropertyManager** to creating the tooling split.
- The core and cavity are added in the **Solid Bodies** folder of **FeatureManager Design Tree**; refer to Figure-40.

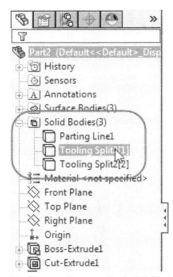

Figure-40. Core and cavity in Solid Bodies folder

Right-click on any of the tooling split and select the **Isolate** option from the menu to check the part separately.

CORE

There are a few portions in the core or cavity that are to be inserted from sides. Such parts are called sliders in engineering. To create sliders, follow the steps given below.

- Isolate the tooling from which you want to extract sliders.
- Click on the **Core** tool from the **Ribbon**.
- Select the side face of the tooling from which you want to create slider. Change the display style to **Hidden Lines Visible** for clear view; refer to Figure-41.

**Face selected for
creating slider**

Figure-41. Face selection for slider

- Create sketch for the slider; refer to Figure-42. You can use offset tool for easy creation of sketch.

Figure-42. Sketch for slider

- Exit the sketch environment and extrude the sketch till the face you want to include in slider.
- Click on the **OK** button from the **PropertyManager** to create the slider; refer to Figure-43.

To move the individual parts of mold, click on the **Move/Copy** option from the **Features** cascading menu of **Insert** menu; refer to Figure-44. The **Move/Copy Body PropertyManager** will display. Select the part that you want to move. Triad will be displayed providing you the options to move the part. Click on the **Translate/Rotate** button at the bottom in the **PropertyManager** if the triad is not displayed. Move the selected part to desired location and click **OK** from the **PropertyManager**. Similarly, you can move other parts of the tool split. Refer to Figure-45.

Slider extacted

Figure-43. Slider created

Figure-44. Move or Copy tool

Figure-45. Tool split parts after moving

PRACTICE 1

Make a mold core and cavity of the model shown in Figure-46 out of a 380x440x100 mm^3 block. The part file is available in resources of the book.

Figure-46. Practice 1 model

SELF ASSESSMENT

Q1. Which of the following analysis is not performed on the part for checking possibility of mold design?

a. Draft Analysis
b. Stress Analysis
c. Undercut Analysis
d. Parting Line Analysis

Q2. If there are steps at parting line, then draft required for shutoff is

a. 1 degree to 3 degree
b. 3 degree to 5 degree
c. 5 degree to 7 degree
d. 7 degree to 9 degree

Q3. To compensate for shrinkage of molded part, we need to use tool.

a. **Draft**
b. **Scale**
c. **Move Face**
d. **Split Line**

Q4. Any opening in the model can be closed by...........tool so that the core steel and cavity steel meet at defined surface.

a. **Shut-off Surfaces**
b. **Parting Surfaces**
c. **Move Face**
d. **Draft**

Q5. ....is the surface by which the core and cavity is separated.

a. Shut-off surface
b. Parting surface
c. Ruled surface
d. Offset surface

FOR STUDENT NOTES

Chapter 13

Sheet metal and Practice

Topics Covered

The major topics covered in this chapter are:

- *Sheet metal Introduction.*
- *Sheet metal Creation Tools.*
- *Sheet metal Design Terms*
- *Tools for applying cut.*
- *Corners Modification*
- *Vents, Rip, and Bend insertion*
- *Flat Pattern.*
- *Solid to Sheet metal*
- *Practice.*

Part   Flat Pattern

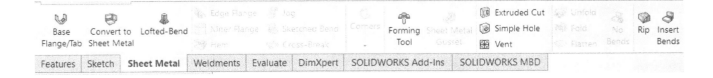

INTRODUCTION TO SHEET METAL

Sheet metal is used when you need a component of thickness in the range of 0.16 mm to 12.70 mm and do not require conventional cutting machines. The components that can be created by Punch-press and bending machines are designed in Sheet Metal environment. In SolidWorks, there is a separate **CommandManager** to design sheet metal components named **Sheet Metal**; refer to Figure-1.

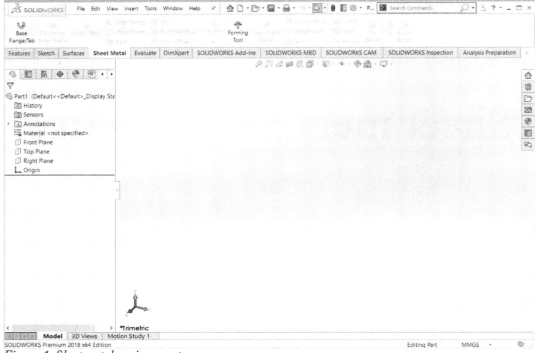

Figure-1. Sheet metal environment

The tools to create sheet metal designs are discussed next.

BASE FLANGE/TAB

The **Base Flange/Tab** tool is used to create base feature of the sheet metal component. All the other features will be created on this base flange. The steps to create base flange/tab are given next.

*   Click on the **Base Flange/Tab** tool. You will be asked to select the sketching plane to draw sketch of the base flange.
*   Select a plane and draw the sketch.
*   Exit the sketch environment. Preview of the flange will be displayed; refer to Figure-2. The **Base Flange PropertyManager** will display; refer to Figure-3 and Figure-4.

Base flange with open sketch | Base flange with closed sketch

Figure-2. Base flange preview

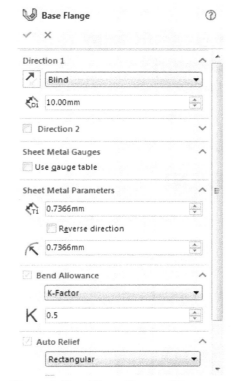

Figure-3. Base Flange PropertyManager with open sketch selected

Figure-4. Base Flange PropertyManager with closed sketch

Setting Parameters for Base Flange/Tab with Open sketch

- Specify the desired length of the flange/tab in direction 1 and direction 2 by using the options in the **Direction 1** and **Direction 2** rollouts in the **PropertyManager**.
- Set the desired thickness and bend radius in the edit boxes available in **Sheet Metal Parameters** rollout or you can use the gauge table for defining parameters.
- To use the gauge table, select the **Use gauge table** check box from the **Sheet Metal Gauges** rollout. You are asked to select a sample table.
- Select the desired sample table from the drop-down; refer to Figure-5.

Figure-5. Sample table selected

- Select the desired gauge of sheet from the **Gauge** drop-down in the **Sheet Metal Parameters** rollout.
- Click on the **OK** button from the **PropertyManager** to create the base flange/tab.

Setting Parameters for Base Flange/Tab with Close sketch

- Set the desired thickness for the flange/tab in the edit box in the **Sheet Metal Parameters** rollout.
- You can also set the parameters by using the gauge table as discussed earlier.

LOFTED-BEND

The **Lofted-Bend** tool is used to create sheet metal component by joining two or more sketch sections. This tool works in the same way as the **Lofted Surfaces** or **Lofted Boss/Base** tool work. Figure-6 shows a lofted bend. Procedure to use this tool is given next.

- Make sure you have two or more open sketches in the graphics area and then click on the **Lofted-Bend** tool from the **Ribbon**. The **Lofted-Bends PropertyManager** will be displayed; refer to Figure-7.

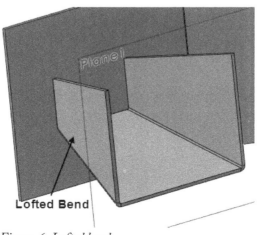
Figure-6. Lofted bend

- Select the **Bent** radio button if you want to create real physical bends, rather than formed geometry and approximated bend lines in a flat pattern. Bent lofted bends form a realistic transition between two profiles to facilitate instructions for press brake manufacturing.

- Select the **Formed** radio button if you want to create a formed transition between two profiles assuming that a forming tool is used to create lofted bend. In this case, you cannot have sharp corners in the profiles; refer to Figure-8.

Figure-8. Formed lofted bend

Figure-7. Lofted-Bends PropertyManager

- In most of the cases, you will be using the **Bent** radio button. Select the **Bent** radio button and then select the profiles.

- Select the **Chord Tolerance** button from the **Faceting Options** rollout and specify the desired tolerance for cord created during bend transition from point to arc of sections; refer to Figure-9. Similarly, you can use the **Number of Bends, Segment Length,** and **Segment Angle** buttons in the same way.

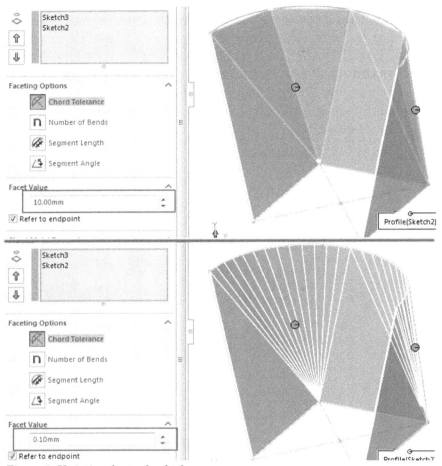

Figure-9. Variation due to chord tolerance

- Specify the sheetmetal parameter in the **Sheet Metal Parameters** rollout as discussed earlier.

SHEET METAL DESIGN TERMS

While going forward in the chapter, you will come across some technical terms are used in sheet metal industry. Here, we will discuss about these technical terms and their effects on production.

Sheet Metal Definition

Metal that has been rolled into a sheet having a thickness between foil and plate. The thickness of metal can vary from fraction of millimeters to 12.5 mm, in general.

Gauge

Gauge is a traditional measurement unit of sheet thickness commonly used in USA, India and many other parts of world. It is a non-linear unit. Higher the gauge number the thinner is the sheet metal. Like gauge 0000000 means 12.7 mm thickness and gauge 38 means 0.16 mm thickness. You can find a table on gauge to mm conversion easily in local stores. Use of Gauge to designate sheet metal thickness is discouraged by numerous international standards organizations. For

Example, ASTM states in specification ASTM A480-10a "The use of gage number is discouraged as being an archaic term of limited usefulness not having general agreement on meaning."

Bend Allowance

When the sheet metal is put through the process of bending the metal around the bend is deformed and stretched. As this happens you gain a small amount of total length in your part. Likewise, when you are trying to develop a flat pattern, you will have to make a deduction from your desired part size to get the correct flat size. The **Bend Allowance** is defined as the material you will add to the actual leg lengths of the part in order to develop a flat pattern. The leg lengths are the part of the flange which is outside of the bend radius. In our example a part with flange lengths of 2" and 3" with an inside radius of .250" at 90° will have leg lengths of 1.625" and 2.625" respectively, refer to Figure-10. When we calculate the Bend Allowance, we find that it equals .457". In order to develop the flat pattern we add .457" to 1.625" and 2.625" to arrive at 4.707". In SolidWorks, you don't need to specify the value of bend allowance as it is automatically calculated based on K factor.

Figure-10. Bend Allowance example

K-Factor

The K-Factor in sheet metal designing is the ratio of the neutral axis to the material thickness. When metal is bent the upper section is going to compress and the lower section is going to stretch. The line where the transition from compression to stretching occurs is called the neutral axis. The location of the neutral axis varies and is based on the material's physical properties and its thickness. The K-Factor is the ratio of the Neutral Axis' Offset (t) and the Material Thickness (MT). Figure-11 shows how the top of the bend is compressed, and the bottom is stretched.

Figure-11. Neutral axis of bent model

$$K = t/MT$$

Generally, K is taken as 0.33 for soft materials and 0.4 for hard materials as a thumb rule. Since, K factor is not just mathematical term, you need experiments to find exact value of K for your material situations. We have given the general steps to find out K-factor by experiment.

Calculating the K-Factor

Since the K-Factor is based on the property of the metal and its thickness, there is no simple way to calculate it ahead of the first bend. Typically the K-Factor is going to be between 0 and .5. In order to find the K-Factor, you will need to bend a sample piece and deduce the Bend Allowance. The Bend Allowance is then plugged into the equation to find the K-Factor.

- Begin by preparing sample blanks which are of equal and known sizes. The blanks should be at least a foot long to ensure an even bend, and a few inches deep to make sure you can sit them against the back stops. For our example let's take a piece that is 14 Gauge, .075", 4" Wide and 12" Long. The length of the piece won't be used in our calculations. Preparing at least 3 samples and taking the average measurements from each will help
- Set up your press brake with the desired tooling you'll be using to fabricate this metal thickness and place a 90° bend in the center of the piece. For our example this means a bend at the 2" mark.
- Once you've bent your sample pieces carefully measure the flange lengths of each piece. Record each length and take the average of lengths. The length should be something over half the original length. For our example the average flange length is 2.073"
- Second measure the inside radius formed during the bending. A set of radius gauges will get you fairly close to finding the correct measurement, however to get an exact measurement an optical comparator will give you the most accurate reading. For our example the inside radius is measured at .105"
- Now that you have your measurements, we'll determine the Bend Allowance. To do this first determine your leg length by subtracting the material thickness and inside radius from the flange length. (Note this equation only works for 90° bends because the leg length

is from the tangent point.) For our example the leg length will be 2.073 - .105 - .075 = 1.893.

- Subtract twice the leg length from the initial length to determine the Bend Allowance. 4 - 1.893 * 2 = .214.
- Plug the Bend Allowance (BA), the Bend Angle (A), Inside Radius (R) and Material Thickness (T) into the below equation to determine the K-Factor (K).

$$K = \frac{-R + \dfrac{BA}{\pi A/180}}{T}$$

EDGE FLANGE

The **Edge Flange** tool is used to create a flange by using selected edge. The steps to do so are given next.

- Click on the **Edge Flange** tool. The **Edge Flange PropertyManager** will display.
- Select an edge using which you want to create the flange wall. The flange end will get attached to cursor; refer to Figure-12.

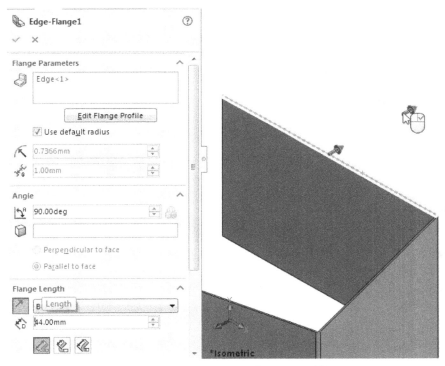

Figure-12. Edge Flange PropertyManager and preview of flange

- Click to specify the end point of the flange.
- You can change the length of the flange by using the spinner in the **Flange Length** rollout.

Some of the important options in the **PropertyManager** are explained next.

Flange Parameters Rollout

The options in the **Flange Parameters** rollout are used to define the edge reference to be used for creating the edge flange, the bending radius, and the profile of the edge flange. These options are discussed next.

Edge

The **Edge selection** box is used to select the edges to create the edge flange.

Edit Flange Profile

The **Edit Flange Profile** button is chosen to edit the profile of the edge flange. By default, the edge flange is created along the entire length of the selected edge. To edit the profile of the edge flange, choose the **Edit Flange Profile** button; the **Profile Sketch** dialog box will be displayed informing you that the sketch is valid. Also, the sketching environment will be invoked in the background. Edit the sketch of the profile of the edge flange using the sketching tools.

You will also notice that while editing the sketch of the edge flange, the **Profile Sketch** dialog box informs you whether the sketch is valid for creating the edge flange or not. If the status of the sketch is shown valid in the **Profile Sketch** dialog box, the preview of the flange will be displayed in the drawing area. You can drag the end points of the line create on selected edge for flange to reduce/increase the width of flange. After editing the profile, choose the **Finish** button from the **Profile Sketch** dialog box; the flange will be created and the **Edge-Flange PropertyManager** will be automatically closed. Note that if you want to modify the other parameters of the flange, choose the **Back** button from the **Profile Sketch** dialog box. Figure-13 shows the edge flange created along the entire length of the selected edge. Figure-14 shows the edited sketch of the edge flange and Figure-15 shows the resulting edge flange.

Angle Rollout

The **Angle** rollout is used to define the angle of the flange. The default angle of the flange is 90 degrees. You can define any other angle of the flange by using the **Flange Angle** spinner. The angle of the edge flange can be greater than 0-degree and less than 180 degrees. You can also select a face and specify whether the resulting flange will be parallel or normal to it. Figure-16 shows an edge flange created at an angle of 45 degrees. Figure-17 shows an edge flange created at an angle of 135 degrees.

Figure-13. Edge flange created along the entire length of the edge

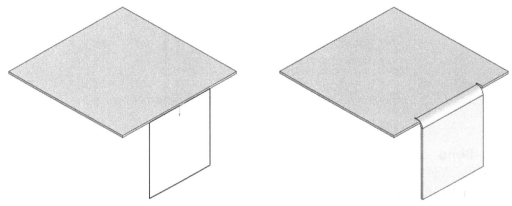

Figure-14. Edited sketch of the edge flange   *Figure-15. Resulting edge flange*

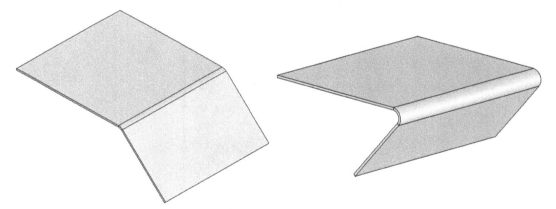

Figure-16. Edge flange created at an angle of 45 degrees   *Figure-17. Edge flange created at an angle of 135 degrees*

Flange Length Rollout

The **Flange Length** rollout is used to define the length of the flange. In other words, the options for feature termination are available in this rollout. These options are the same as discussed earlier. The other two options provided in this rollout are discussed next.

Outer Virtual Sharp

The **Outer Virtual Sharp** button is used to define the length of the flange from the outer virtual sharp. The outer virtual sharp is an imaginary vertex created by extending the tangent lines virtually from the outer radius of the bend, as shown in Figure-18.

Inner Virtual Sharp

The **Inner Virtual Sharp** button is chosen by default and is used to define the length of the flange from the inner virtual sharp. The inner virtual sharp is an imaginary vertex created by extending the tangent lines virtually from the inner radius of the bend, as shown in Figure-18.

Figure-18. Outer Virtual Sharp and Inner Virtual Sharp

Tangent Bend

The **Tangent Bend** button is used to define the length of the flange from the imaginary line that is created by extending the tangent line from the outer radius of the bend and parallel to the end edge of the flange to be created, refer to Figure-19. This button will be available only for the flange to be created whose bend radius is greater than 90-degree.

Figure-19. Tangent length of
Tangent Bend *in base flange*

Flange Position Rollout

The **Flange Position** rollout is used to define the position of the flange on an edge. The options in this rollout are discussed next.

Material Inside

The **Material Inside** button is used to create the edge flange in such a way that the material of the flange after the bend lies inside the maximum limit of sheet. Figure-20 shows the edge flange created with the Material Inside button chosen.

Material Outside

The **Material Outside** button is chosen by default and the edge flange is created such that the material of the flange after the bend lies outside the maximum limit of the sheet. Figure-21 shows the edge flange created with the **Material Outside** button chosen.

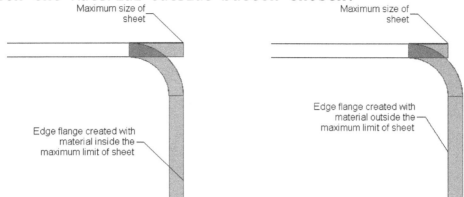

Figure-20. Edge flange created with the ***Material Inside*** *button chosen*

Figure-21. Edge flange created with the ***Material Outside*** *button chosen*

Bend Outside

The **Bend Outside** button is used to create an edge flange such that the bending of the sheet starts from the point that is beyond the maximum limit of the sheet, as shown in Figure-22.

Bend from Virtual Sharp

The **Bend from Virtual Sharp** button is used to create an edge flange with the bending of the sheet starting from the virtual sharp. The position of the flange depends on whether you choose the **Outer Virtual Sharp** button, **Inner Virtual Sharp** button, or the **Tangent Bend** button from the **Flange Length** rollout. Figure-23 shows the edge flange created with the **Inner Virtual Sharp** and **Bend from Virtual Sharp** buttons chosen.

Figure-22. Edge flange created with the ***Bend Outside*** *button chosen*

Figure-23. Edge flange created with the ***Bend from Virtual Sharp*** *button chosen*

Tangent to Bend

The **Tangent to Bend** button is used to create the edge flange in such a way that the material of the flange after bending lies tangent to the maximum limit of the sheet, refer to Figure-24. Note that this option is not valid for the bend angle less than 90 degree.

Figure-24. Edge flange created with the **Tangent to Bend** *button chosen*

Trim side bends

Select the **Trim side bends** check box to trim extra materials in the bends surrounding the current edge flange. By default, this check box is not selected. Figure-25 shows the edge flange created with the **Trim side bends** check box cleared. Figure-26 shows the edge flange created with the **Trim side bends** check box selected.

Figure-25. Edge flange created with the **Trim side bends** *check box cleared*

Figure-26. Edge flange created with the **Trim side bends** *check box selected*

Offset

The **Offset** check box is available only when you create an edge flange using the **Material Inside, Material Outside, Bend Outside,** or **Tangent to Bend** options. This check box is used to create an edge flange at an offset distance from the selected edge reference. On selecting the **Offset** check box, the **Offset End Condition** drop-down and the **Offset Distance** spinner will be displayed. Specify the offset distance using the spinner. Figure-27 shows the edge flange created with the **Offset** check box cleared. Figure-28 shows the edge flange created with the **Offset** check box selected and the offset distance specified in the **Offset Distance** spinner.

*Figure-27. Edge flange created with the
Offset check box cleared*

*Figure-28. Edge flange created with the
Offset check box selected*

Custom Bend Allowance Rollout

The **Custom Bend Allowance** rollout is used to define the bend allowance other than the default bend allowance that you defined while creating the base flange. To apply the custom bend allowance, expand this rollout by selecting the **Custom Bend Allowance** check box. Then use the options in this rollout to define the bend allowance for the current bend as discussed earlier.

Custom Relief Type Rollout

The **Custom Relief Type** rollout is used to define the type of relief other than the default that was defined while creating the base flange. To apply the custom relief, expand this rollout by selecting the check box in the title bar of the **Custom Relief Type** rollout, as shown in Figure-29.

The types of reliefs that can be defined for a sheet metal component are discussed next.

Obround Relief

The **Obround** option is used to provide the obround relief such that the edges of the relief merging with the sheet are rounded. The **Use relief ratio** check box is selected by default. Therefore, you can modify the value of the relief ratio by setting the value in the **Relief Ratio** spinner. If you clear the **Use relief ratio** check box, the **Relief Width** and **Relief Depth** spinners will be displayed, as shown in Figure-30. You can modify the relief width and relief depth individually by using these two spinners.

Figure-31 shows the edge flange created by providing the obround relief with the default relief ratio. Figure-32 shows the edge flange created by providing obround relief after modifying the relief ratio.

*Figure-29. The **Custom Relief Type** rollout*

*Figure-30. The **Relief Width** and **Relief Depth** spinners displayed in the **Custom Relief Type** rollout*

Figure-31. Edge flange created with the default relief ratio

Figure-32. Edge flange created after modifying the relief ratio

Rectangle Relief

The **Rectangle** option is selected by default in this rollout. This option is used to provide the rectangular relief to the sheet metal components. The options for defining the rectangular relief are the same as discussed in the previous paragraph. Figure-33 shows an edge flange created by providing the rectangular relief with the default relief ratio. Figure-34 shows an edge flange created by providing rectangular relief after modifying the relief ratio.

Figure-33. Edge flange created by providing the rectangular relief with default relief ratio

Figure-34. Edge flange created by providing the rect-angular relief after modifying the relief ratio

Tear Relief

You can provide the tear relief to an edge flange by using the **Tear** option. The tear relief will tear the sheet in order to accommodate the bending of the sheet. When you select the **Tear** option from the **Relief Type** drop-down list, all the other options are replaced by the **Rip** and **Extend** buttons, as shown in Figure-35.

*Figure-35. The **Custom Relief Type** rollout with the **Tear** option selected from the **Relief Type** drop-down list*

The **Rip** button is chosen by default. This option rips or tears the sheet to accommodate the bending of the sheet, as shown in Figure-36. When the **Extend** button is chosen, the outer faces of the bend will be extended to the outer faces of the sheet on which you create the edge flange, as shown in Figure-37.

*Figure-36. Tear relief with the **Rip** button chosen*   *Figure-37. Tear relief with the **Extend** button chosen*

MITER FLANGE

The **Miter Flange** tool is used to create a flange of specified shape. This type of flange is best used in creating tray type shapes. The steps to create miter flange are given next.

- Click on the **Miter Flange** tool from the **Ribbon**. You are asked to select a plane/edge to define sketching plane.
- Select the edge on which you want to create flange. The sketching environment will display.
- Create an open sketch to define shape of flange; refer to Figure-38 and exit the sketching environment. Preview of flange will display; refer to Figure-39.

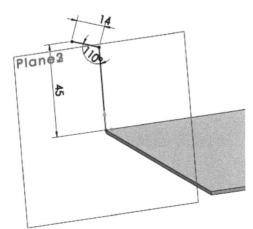

Figure-38. Sketch for miter flange

Figure-39. Preview of miter flange

- If you want to create tray then select all the edges of the current loop; refer to Figure-40.

Figure-40. Closed loop of miter flange

- Click on the **OK** button to create the flange.

Some of the important options in the **PropertyManager** are explained next.

Gap distance Area

The spinner in this area is used to define the rip distance between two consecutive flanges. Set the value in the **Rip Gap** spinner to modify the distance value of the rip. Figure-41 shows the miter flange created using the default distance value and Figure-42 shows the miter flange created using the modified rip distance. In case of error enter the higher gap value.

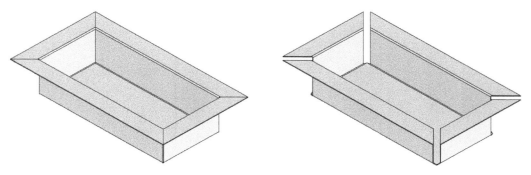

Figure-41. Miter flange with the default rip distance *Figure-42. Miter flange with the modified rip distance*

Start/End Offset Rollout

You can specify the start and end offset distances of the miter flange by using the options in the **Start/End Offset** rollout. The **Start Offset Distance** spinner is used to specify the offset distance from the start face of the miter flange. The **End Offset Distance** spinner is used to specify the offset distance from the end face of the miter flange. If the start and end offset distances are applied to the miter flange created on the continuous edges of the base flange, the start offset distance will be applied to the first edge and the end offset distance will be applied to the edge selected at last. Figure-43 shows the miter flange created on a single edge with the start and end offsets. Figure-44shows the offsets applied to the miter flange created by selecting all the edges of the base flange.

Figure-43. Miter flange created at an offset distance on a single edge *Figure-44. Miter flange created at an offset distance on all edges*

HEM

The **Hem** tool is used to create bend at the end edge of the sheet. The steps to create hems are given next.

• Click on the **Hem** tool from the **Ribbon**. The **Hem PropertyManager** will display; refer to Figure-45.

Figure-45. Hem PropertyManager

- Select the edge on which you want to create the hem. Preview of hem will display; refer to Figure-46.

Figure-46. Preview of hem

- Select the desired shape using the **Closed, Open, Tear Drop,** or **Rolled** button from the **Type and Size** rollout.
- Specify the size and other parameters and click on the **OK** button from the **PropertyManager** to create the hem.

JOG

The **Jog** tool can be used to create double bend in the sheet. The steps to create jog are given next.

- Click on the **Jog** tool from the **Ribbon**. You are asked to select a plane for creating sketch.
- Sketch the bend line on the sheet metal face and exit the sketching environment. You are asked to select a face that you want to be fixed.
- Select the fixed face. The preview of jog will display; refer to Figure-47.

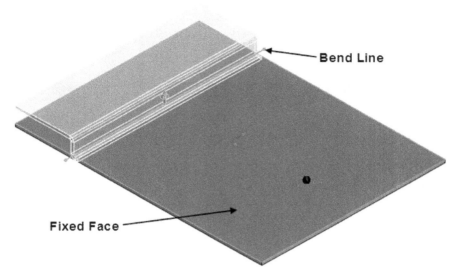

Figure-47. Preview of Jog

- Specify the desired parameters in the **PropertyManager** displayed and click on the **OK** button to create the bend.

SKETCHED BEND

The **Sketched Bend** tool is used to bend a sheet metal part by specified angle at selected bend line. The steps to bend a part are given next.

- Click on the **Sketched Bend** tool from the **Ribbon**. You are asked to select a flat face to bend.
- Select the face of sheet metal part. The sketching environment will display.
- Draw the bend line in such a way that it intersect to the edges of the sheet metal part's face.
- Exit the sketch and specify the parameters in the **PropertyManager**.
- Click on the **OK** button from the **PropertyManager**. The bend will be created with specified parameters; refer to Figure-48.

Figure-48. Sketched Bend

CROSS-BREAK

The **Cross-Break** tool is used to create cross-break in the sheet metal part. In HVAC or other duct works, cross-breaks are provided to stiffen the sheet. The procedure to use this tool is given next.

- Click on the **Cross-Break** tool from the **Sheet Metal** tab in the **Ribbon**. The **Cross Break PropertyManager** will be displayed; refer to Figure-49.
- Select the face on which you want to apply the cross-break. Preview of the cross-break will be displayed.
- Set the desired parameters and click on the **OK** button from the **PropertyManager**.

CLOSE CORNER

The **Close Corner** tool is used to change the closing of flange walls at corners. The steps to use this tool are given next.

- Click on the **Closed Corner** tool from **Corners** drop-down in the **Ribbon**. The **Closed Corner PropertyManager** will display.
- Select the face at corner that you want to close. Preview will be displayed; refer to Figure-50.

Figure-49. Cross Break Property-Manager

Figure-50. Preview of closed corner

- Set the desired options in the **PropertyManager** and click on the **OK** button to create closed corner.

WELDED CORNER

The **Welded Corner** tool is used to weld the gap between two adjacent walls. The procedure to use this tool is similar to **Close Corner** tool. The procedure is discussed next.

- Click on the **Welded Corner** tool from the **Corners** drop-down in the **Ribbon**; refer to Figure-51. The **Welded Corner PropertyManager** will be displayed; refer to Figure-52.

Figure-51. Corner drop-down

Figure-52. Welded Corner Property-Manager

- Select the side face of sheet metal wall to be closed by welding bead. Preview of the welded corner will be displayed; refer to Figure-53.

Figure-53. Preview of welded corner

- Set the fillet radius and other parameters in the **PropertyManager** and click on the **OK** button to create the welded corner.

BREAK-CORNER/CORNER-TRIM

The **Break-Corner/Corner-Trim** tool is used to chip-off the sharp corners of the sheet metal component. The procedure to use this tool is given next.

- Click on the **Break-Corner/Corner-Trim** tool from the **Corners** drop-down in the **Ribbon**. The **Break Corner PropertyManager** will be displayed; refer to Figure-54.
- Select the sharp edge of the sheet metal component. Preview of break corner will be displayed; refer to Figure-55.

Figure-54. Break Corner Prop-ertyManager

Figure-55. Preview of break corner

- Set the desired parameters and click on the **OK** button from the **PropertyManager** to create the corner break.

CORNER RELIEF

The **Corner Relief** tool is used to apply bend relief at the corners of sheet metal parts. The procedure to use this tool is given next.

- Click on the **Corner Relief** tool from the **Corners** drop-down in the **Ribbon**. The **Corner Relief PropertyManager** will be displayed; refer to Figure-56.
- Select the **2 Bend Corner** or **3 Bend Corner** radio button from the **Corner Type** rollout of the **PropertyManager** to modify the respective type of corners.
- Click on the **Collect all corners** button from the **PropertyManager**. All the corners in the sheet metal model will be displayed in the selection box in **PropertyManager**.
- Select one or all the corners from the selection box and select the desired button from the **Relief Options** rollout in the **PropertyManager**. The respective type of corner relief will be applied to the sheet metal body.
- Click on the **OK** button from the **PropertyManager** to apply selected corner relief.

SHEET METAL GUSSET

The **Sheet Metal Gusset** tool is used to add rib to support any sheet metal flange. The steps to create sheet metal gusset are given next.

Figure-56. Corner Relief Property-Manager

- Click on the **Sheet Metal Gusset** tool from the **Ribbon**. The **Sheet Metal Gusset PropertyManager** will display; refer to Figure-57.
- Select the two faces (one of base and other of flange). The preview of gusset will display; refer to Figure-58.
- Set the desired parameters and click on the **OK** button from the **PropertyManager** to create the gusset.

Figure-57. Sheet Metal Gusset
PropertyManager

Figure-58. Preview of gusset

EXTRUDE CUT

The **Extrude Cut** tool is used to create cuts in the sheet metal part by extrusion. The steps to create extruded cut are given next.

- Click on the **Extruded Cut** tool from the **Ribbon**. You are asked to select the face of sheet metal part.
- Click on the face and draw the closed sketch of the cut.
- Exit the sketch and click on the **OK** button to create the cut.

VENT

The **Vent** tool is used to create vents in the sheet metal part for air/ material circulation. The procedure to use this tool is given next.

- Create a sketch for vent feature on the sheet metal face which should have geometry for boundary, ribs, spars, and fill-in boundary; refer to Figure-59.

Figure-59. Sketch for vent feature

- Click on the **Vent** tool from the **Sheet Metal CommandManager** in the **Ribbon**. The **Vent PropertyManager** will be displayed; refer to Figure-60. Also, you will be asked to select the entities to be used as boundary of vent.

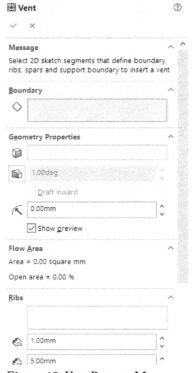

Figure-60. Vent PropertyManager

- Select the closed sketch entity to be used as boundary of vent.
- Set the radius and draft values in the **Geometry Properties** rollout if you wish to.
- Scroll-down in the **PropertyManager** and click in the **Ribs** selection box. Click on the open sketches to create rib of vent; refer to Figure-61.

Figure-61. Preview of vent ribs

- Set the width of rib in the edit box of **Ribs** rollout in **PropertyManager**.
- Click in the **Spars** selection box and select the sketch(s) for spars. Preview will be displayed. Set the desired width of spars.
- Click in the **Fill-In Boundary** selection box and select the closed region created for fill-in boundary; refer to Figure-62.

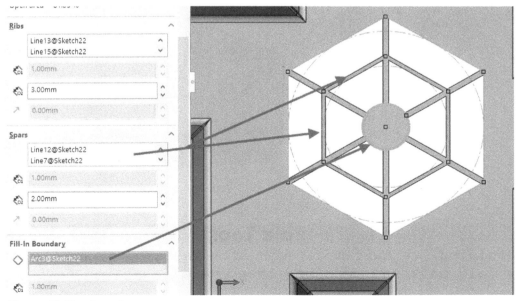

Figure-62. Preview of vent

- Set the desired parameters and click on the **OK** button to create the vent.

UNFOLD TOOL AND FOLD TOOL

The **Unfold** tool is used to unbend the selected bends. While the Fold tool does the reverse of unfold tool. After unbending you can perform editing which will be reflected in the folded part. In most of the cases, the **Unfold** and **Fold** tools work together. The procedure to use unfold tool is given next.

Unfold Tool

• Click on the **Unfold** tool from the **Sheet Metal CommandManager** in the **Ribbon**. The **Unfold PropertyManager** will be displayed as shown in Figure-63.

Figure-63. Unfold Property-Manager

• Select the face that you want to be fixed while unfolding. You will be asked to select the bends to be unfolded.
• Select the bends and then click on the **OK** button; refer to Figure-64. The sheet metal part will be unfolded.

Figure-64. Selection for unfolding

Fold Tool

After unfolding, we have made some cuts in the part at the bends. We will now fold back the unfolded sheet metal as given next.

• Click on the **Fold** tool from the **Sheet Metal CommandManager** in the **Ribbon**. The **Fold PropertyManager** will be displayed; refer to Figure-65.

Figure-65. Fold Property-Manager

- Select the fixed face if not selected by default and then select the unfolded bends.
- Click on the **OK** button from the **PropertyManager** to fold the unfolded bends; refer to Figure-66.

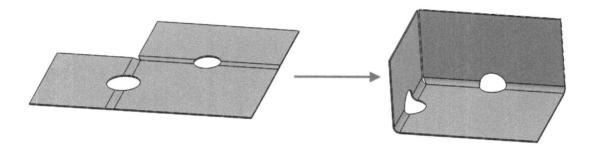

Figure-66. Folding unfolded bends

CONVERT TO SHEET METAL

The **Convert to Sheet Metal** tool is used to convert a solid body into sheet metal component. The procedure to use this tool is given next.

- Make sure you have a solid object in the drawing area and then click on the **Convert to Sheet Metal** tool from the **Sheet Metal CommandManager** in the **Ribbon**. The **Convert to Sheet Metal PropertyManager** will be displayed; refer to Figure-67. Also, you will be asked to select the base of sheet metal part.
- Select a flat face of the model. Specify the sheet thickness and default bend radius in the respective spinners in **Sheet Metal Parameters** rollout of the **PropertyManager**; refer to Figure-68.
- Click in the **Bend Edges** selection box and select the edges of the model where bends are to be applied; refer to Figure-69.

Figure-67. Convert To Sheet Metal PropertyManager

Figure-68. Selection for Sheet Metal Parameters

Figure-69. Edges selected for conversion in bends

- Scroll-down in the **PropertyManager** and specify the desired parameters for corners, bend allowance, and auto relief.
- Click on the **OK** button from the **PropertyManager** to convert the solid into Sheet Metal part; refer to Figure-70.

Figure-70. Solid to sheet metal conversion

RIP TOOL

The **Rip** tool is used to remove sharp edges and replace then with specified gap. This tool is mainly used when you need to convert solid parts into sheet metal. The procedure to use this tool is given next.

- Click on the **Rip** tool from the **Sheet Metal CommandManager**. The **Rip PropertyManager** will be displayed; refer to Figure-71.

Figure-71. Rip PropertyManager

- Select the edge(s) of the model that you want to be ripped. Set the gap size in the **Rip Gap** edit box of the **PropertyManager**.
- Click on the **OK** button from the **PropertyManager** to create the rip feature; refer to Figure-72.

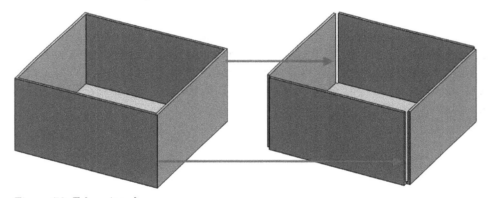

Figure-72. Edges ripped

INSERT BENDS TOOL

The **Insert Bends** tool is used to insert bends in the solid model so that it can be used in sheet metal work. The procedure to use this tool is given next.

- Click on the **Insert Bends** tool from the **Sheet Metal CommandManager**. The **Bends PropertyManager** will be displayed; refer to Figure-73. Also, you will be asked to select the fixed face of the model.
- Select the flat face that you want to be fixed.
- Specify the parameters like bend radius, bend allowance, relief etc. in **PropertyManager**.
- Click in the **Edges to Rip** selection box in the **Rip Parameters** rollout of the **PropertyManager** and select the edges that you want to be ripped.

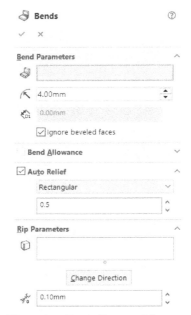

Figure-73. Bends PropertyManager

- Click on the **OK** button from the **PropertyManager**. Bends will be applied automatically to the sharp edges.

FLATTEN

The **Flatten** tool is used to create flat pattern of the sheet metal part. Click on the tool from the **Ribbon** and Flat pattern will be displayed. In some cases, you may be asked to specify the base. In those cases, select the flat face that you want to be fixed. Figure-74 shows a sheet metal part and its flat pattern.

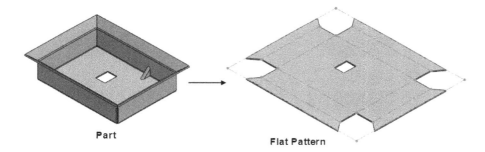

Figure-74. Part and its flat pattern

INSERTING FLAT PATTERN IN DRAWING

- Start the drawing by using the **Make Drawing from Part/Assembly** button from the **New** drop-down of **Menu Bar**; refer to Figure-75. The drawing mode will open.

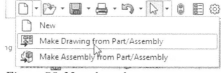

Figure-75. New drop-down

- Select the sheet size from the list and click on the **OK** button from the dialog box displayed. The **View Palette** will be displayed in the right of the drawing; refer to Figure-76.

Figure-76. View Palette

- Drag the **Flat pattern** view from the palette and place it at desired location in the drawing; refer to Figure-77.

Figure-77. Flat pattern view

PRACTICE

Create the sheet metal model as shown in Figure-78. Dimensions are given in Figure-79.

Figure-78. Sheet metal model

Thickness of sheet= 1mm
Bend Radius= 5mm

Figure-79. *Drawing views*

To get more drawings for practice, write us at **cadcamcaeworks@gmail.com**

SELF ASSESSMENT

Q1. The...........tool is used to create base feature of the sheet metal component.

Q2. The.............tool is used to create sheet metal component by joining two or more sketch sections.

Q3. The............relief will tear the sheet in order to accommodate the bending of the sheet.

Q4. The.............tool is used to create a flange of desired shape.

Q5. The............tool is used to create bend at the end edge of the sheet.

Q6. The **Jog** tool can be used to create double bend in the sheet. (T/F)

Q7. In HVAC or other duct works, cross-breaks are provided to stiffen the sheet. (T/F)

Q8. The **Close Corner** tool is used to make all the corners of a sheet metal part closer. (T/F)

FOR STUDENT NOTES

FOR STUDENT NOTES

Chapter 14

Weldments

Topics Covered

The major topics covered in this chapter are:

- *Introduction.*
- *Weldment tool.*
- *Structural members.*
- *End Cap.*
- *Weld Beads.*
- *Welding Symbols in Drawing*

ITEM NO.	QTY.	DESCRIPTION	LENGTH
1	1	PIPE, SCH 40, 0.50 DIA.	6.09
2	1	PIPE, SCH 40, 0.50 DIA.	5.58
3	1	PIPE, SCH 40, 0.50 DIA.	22.02
4	1	PIPE, SCH 40, 0.50 DIA.	5.6

INTRODUCTION

This chapter is dedicated to welding joints or so called weldments. Welding is a method to permanently join parts with the help of welding beads. In SolidWorks, we can display the welding beads in model as well as in drawing. In model, we can display a solid bead of weld around the selected faces/edges. We can also attach the welding bead symbol there. In drawing, we can attach the welding symbol to the affected edges/faces. We can also insert the cut list to tabulate the components used in the model. But, before we start using SolidWorks for weldments, its important to revise some basics of welding.

WELDING SYMBOLS AND
REPRESENTATION IN DRAWING

The symbols to represent various type of welds are given next.

Butt/Groove Weld Symbols

Various symbols that come under this category are given next. Refer to Figure-1 and Figure-2.

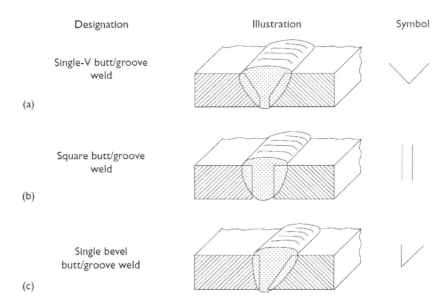

Figure-1. Welding symbols list 1

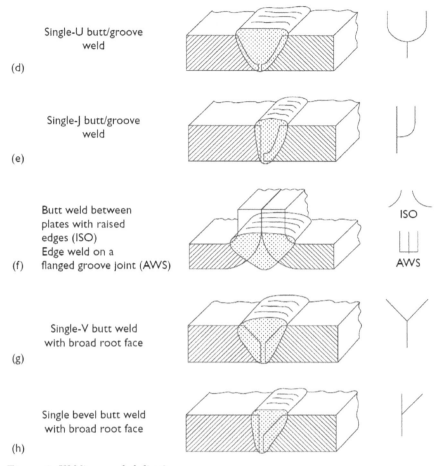

(d) Single-U butt/groove weld

(e) Single-J butt/groove weld

(f) Butt weld between plates with raised edges (ISO)
Edge weld on a flanged groove joint (AWS)

ISO

AWS

(g) Single-V butt weld with broad root face

(h) Single bevel butt weld with broad root face

Figure-2. Welding symbols list 2

Fillet and Edge Weld Symbols

Various symbols that come under this category are given next. Refer to Figure-3.

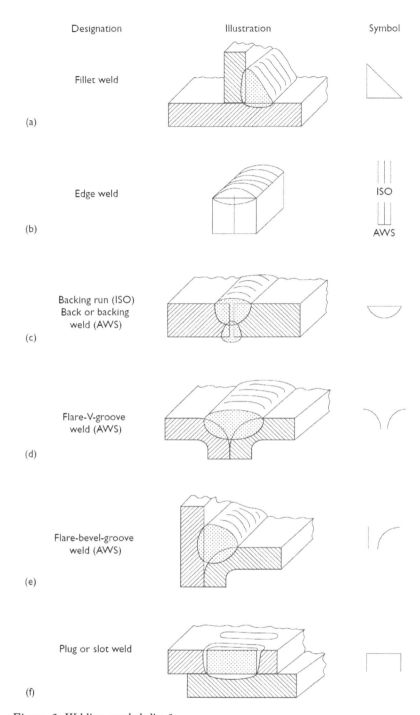

Figure-3. Welding symbols list 3

Miscellaneous Weld Symbols

Various symbols that come under this category are given next. Refer to Figure-4.

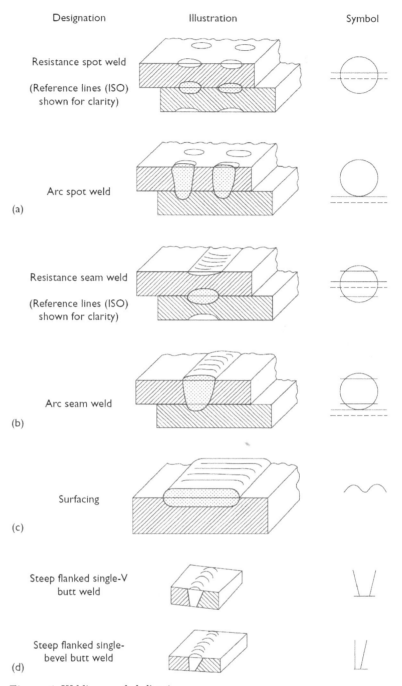

Designation	Illustration	Symbol

Figure-4. Welding symbols list 4

Now, we know various symbols used in welding drawings but keep a note that placement of welding symbol along the arrow decides the side on which the welding will be done on the object; refer to Figure-5.

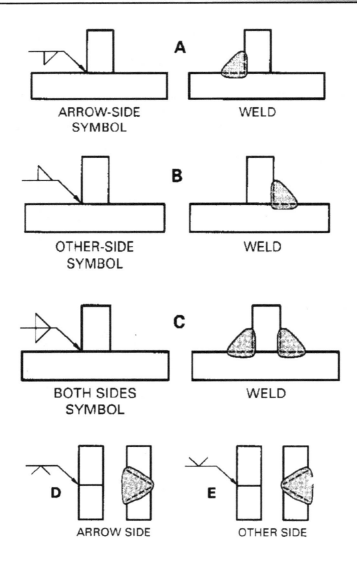

Figure-5. Deciding weld bead side

Dimensioning a weld bead

Alike the other measurements, weld is also measured with respect to various references so that we can control its quality. Figure-6 shows the information required for dimensioning a weld bead.

Figure-6. Welding dimension

Till this point, we have learned the basics of weld symbol representations in drawings. So, we are ready to dive into SolidWorks for creating welding representations.

The tools to add weldments are available in the **Weldments CommandManager** of the **Ribbon**; refer to Figure-7. If this **CommandManager** is not displayed by default, then you can add it by right-clicking on any of the **CommandManager** in the **Ribbon** and then selecting the **Weldments** option from the shortcut menu displayed; refer to Figure-8.

Figure-7. Weldments CommandManager

Figure-8. Weldments option in shortcut menu

Most of the tools in this tab have been discussed earlier. The remaining tools are discussed next.

WELDMENT TOOL

The **Weldment** tool is used to start multi-body environment in which you can create parts without merging them into a single body. In this way, we are allowed to join two entities in the modeling environment with the help of weld beads. The **Weldment** tool does not do a function like **Extrude** or **Revolve** tool but it starts **Weldment** environment. Click on this tool to start the environment.

STRUCTURAL MEMBER

The **Structural Member** tool is used to create structural member for welding, like angle, C channel, pipe, and so on. The procedure to use this tool is given next.

- Click on the **Structural Member** tool from the **Ribbon**. The **Structural Member PropertyManager** will be displayed; refer to Figure-9.

Figure-9. Structural Member PropertyManager

- Select the desired standard, type and size of structural member from the **PropertyManager**.
- One by one click on the line member of the sketch. Preview of the structural member will be displayed; refer to Figure-10.
- Set the desired parameters and click on the **OK** button from the **PropertyManager** to create the structure.

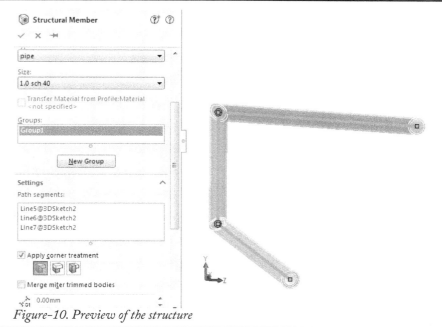

Figure-10. Preview of the structure

END CAP TOOL

The **End Cap** tool is used to close the ends of an open structural member by using a lid. The procedure to use this tool is given next.

- Click on the **End Cap** tool from the **Ribbon**. The **End Cap PropertyManager** will be displayed; refer to Figure-11.

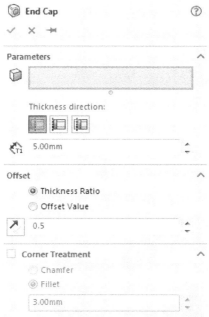

Figure-11. End Cap PropertyManager

- Select the end face of the structural member. Preview of the end cap will be displayed; refer to Figure-12.
- Select the desired thickness direction using the three buttons in the **Thickness direction:** area of the **Parameters** rollout.
- Specify the desired thickness in the **Thickness** edit box.
- Specify the other required parameters and then click on the **OK** button from the **PropertyManager** to create the end cap.

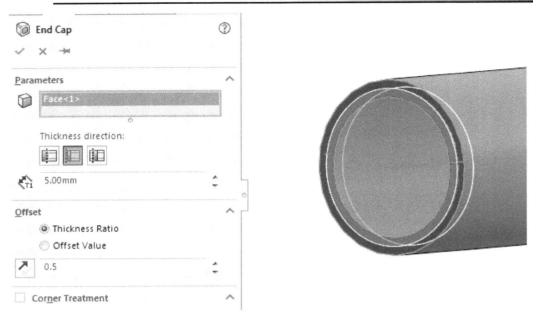

Figure-12. Preview of the end cap

WELD BEAD

The **Weld Bead** tool is used to apply the desired type of welding bead on the selected edges. The procedure to use this tool is given next.

- Click on the **Weld Bead** tool from the **Ribbon**. The **Weld Bead PropertyManager** will be displayed; refer to Figure-13.

Figure-13. Weld Bead PropertyManager

- If **Weld Geometry** radio button is selected in the **Weld selection** area of the **Settings** rollout then select the face sets of the two joining parts; refer to Figure-14.

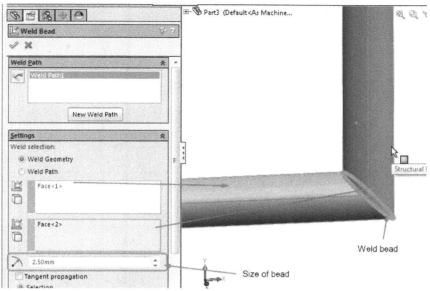

Figure-14. Weld bead using faces

- If you select the **Weld Path** radio button from the **Weld selection** area of the **Settings** rollout then you need to select the edge/edges along which you want to represent welding bead; refer to Figure-15.

Figure-15. Weld bead using edge

- After creating every closed bead, you need to start a new weld path by clicking on the **New Weld Path** tool from the **Weld Path** rollout in the **PropertyManager**.
- To display the weld symbol on the model, click on the **Define Weld Symbol** button from the **Settings** rollout. A dialog box will be displayed; refer to Figure-16.

Figure-16. Weld Symbol dialog box

- Now, this is the dialog box in which you need to feed all the information regarding the weld bead. You can go back to the starting of this chapter for symbol reference and then specify the dimensions for weld bead.
- Click on the **Weld Symbol** button in the dialog box and select the desired welding symbol; refer to Figure-17. Click on the **More Symbols** option from the flyout if desired symbol is not listed.

Figure-17. Weld symbols

- After specifying the desired dimension, click on the **OK** button from the dialog box.
- To set the desired length of weld bead, select the **From/To Length** check box and specify the desired length of bead; refer to Figure-18.
- Similarly, you can select the **Intermittent Weld** check box and specify the intermediate gap for welding bead; refer to Figure-19.

Figure-18. Setting length of weld bead

Figure-19. Intermittent weld

- After specifying the desired parameters, click on the **OK** button from the **PropertyManager** to create the bead. The bead will be added in the **Weld Folder** in **FeatureManager Design Tree**; refer to Figure-20.

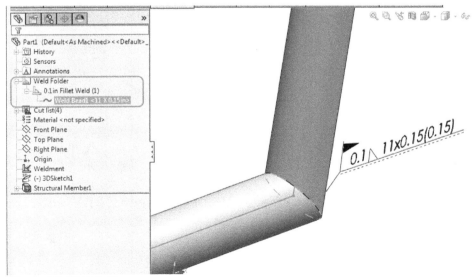

Figure-20. Weld Folder

INSERTING WELDING DATA IN DRAWING

There is no use of assigning welding symbols unless the manufacturer/ fabricator does not get them in his drawing. The procedure to insert the welding data in the drawing is given next.

- After creating the welding model, click on the **Make Drawing from Part** option from the **File** menu; refer to Figure-21. The drawing environment will be displayed.

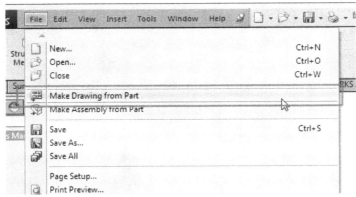

Figure-21. Make Drawing from Part option

- Select the desired sheet size and create the basic views; refer to Figure-22.
- Select the view/views in which you want to display the annotations and click on the **Model Items** button from the **Annotations** tab in the **Ribbon**. The **Model Items PropertyManager** will be displayed.
- Select the **Weld Symbols** button from the **Annotations** rollout in the **PropertyManager**; refer to Figure-23.

Figure-22. Drawing from part

Figure-23. Weld Symbols in Model Items PropertyManager

- Click on the **OK** button from the **PropertyManager**. The welding symbols will be assigned in the drawing; refer to Figure-24.

Figure-24. Drawing with welding symbols

Inserting the Cut list

Cut list is a kind of part list similar to bill of material. The cut list is used to identify the structural members that are being joined by using the welding bead. The procedure to insert the cut list is given next.

- Click on the **Weldment Cut List** option from the **Tables** drop-down in the **Annotation CommandManager** of the **Ribbon**; refer to Figure-25. You are asked to select a view for which you want to create the cut list.
- Select the desired view. The **Weldment Cut List PropertyManager** will be displayed; refer to Figure-26.

Figure-25. Weldment cut list option

Figure-26. Weldment Cut List PropertyManager

- Specify the desired options, if any. Next, click on the **OK** button from the **PropertyManager**. The cut list will get attached to the cursor; refer to Figure-27.

ITEM NO.	QTY.	DESCRIPTION	LENGTH
1	1	PIPE, SCH 40, 0.50 DIA.	6.09
2	1	PIPE, SCH 40, 0.50 DIA.	5.58
3	1	PIPE, SCH 40, 0.50 DIA.	22.02
4	1	PIPE, SCH 40, 0.50 DIA.	5.6

Figure-27. Cut list attached to the cursor

- Click at the desired location to place the cut list. Similarly, you can place the Weld table to specify the welding length and parameters.

SELF ASSESSMENT

Q1. Draw the symbol of Square butt/groove weld.

Q2. Draw the symbol of Single bevel butt weld with broad root face.

Q3. Draw the symbol of Fillet weld.

Q4. Draw the symbol of Plug weld.

Q5. Draw the symbol of Single-J butt/groove weld.

Q6. Which of the following figure shows correct annotation of welding symbol?

a.

b.

c. Both of the above.
d. None of the above.

Q7. Which of the following is not a type of structural member in Structural Member PropertyManager?

a. pipe
b. round bar
c. c channel
d. s section.

FOR STUDENT NOTES

FOR STUDENT NOTES

Chapter 15

3D Printing and
Model Based Definition
(MBD)

Topics Covered

The major topics covered in this chapter are:

- ***3D Printing and Processes***
- ***Print 3D Tool***
- ***SolidWorks MBD (Model Based Definition)***
- ***DimXpert Tools***
- ***3D PDF generation***
- ***3D PMI Compare***

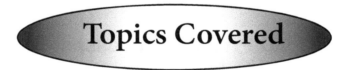

Auto Dimension | Location | Size | Basic Location Dimension | Datum | Geometric | Pattern | Copy | Show Tolerance | Datum | Surface | Weld | Note | Smart | Tables | Capture | Section | Model | Exploded
Scheme | Dimension | Dimension | | | Tolerance | Feature | Scheme | Status | Target | Finish | Symbol | | Dimension | | 3D View | View | Break View | View

Features | Sketch | Weldments | Data Migration | Direct Editing | Evaluate | DimXpert | SOLIDWORKS Add-Ins | **SOLIDWORKS MBD**

3D PRINTING

3D Printing also known as Additive Manufacturing is not a new concept as it was developed in 1981. Since then 3D Printing technology is continuously evolving. In early stages, the 3D printers were able to create only prototypes of objects using the polymers. But now a days, 3D printers are able to produce final products using metals, plastics and biological materials. 3D printers are being used for making artificial organs, architectural art pieces, complex design objects etc. Although, 3D Printing technique was created for manufacturing industry but now, it has found more applications in medical field.

In SolidWorks, there is a very simple and robust mechanism for 3D printing. The procedure of 3D Printing itself is not difficult but it is important to prepare your part well for 3D printing. We will first discuss the part preparation for 3D Printing and then we will use SolidWorks tools for performing 3D print.

PART PREPARATION FOR 3D PRINTING

Part preparation is very important step for 3D Printing. If your part is not stable in semi molten state then it is less suitable for 3D printing. Stability of model is directly dependent on the material you are using for 3D Printing. We will know more about part preparation but before that it is important to understand different type of processes available in 3D Printing.

3D Printing Processes

Not all 3D printers use the same technology. There are several ways to print and all those available are additive, differing mainly in the way layers are build to create the final object.
Some methods use melting or softening material to produce the layers. Selective laser sintering (SLS) and fused deposition modeling (FDM) are the most common technologies using this way of 3D printing. Another method is when we talk about curing a photo-reactive resin with a UV laser or another similar power source one layer at a time. The most common technology using this method is called stereolithography (SLA).

In 2010, the American Society for Testing and Materials (ASTM) group "ASTM F42 - Additive Manufacturing", developed a set of standards that classify the Additive Manufacturing processes into 7 categories according to Standard Terminology for Additive Manufacturing Technologies. These seven processes are:

1. Vat Photopolymerisation
2. Material Jetting
3. Binder Jetting
4. Material Extrusion
5. Powder Bed Fusion
6. Sheet Lamination
7. Directed Energy Deposition

Brief introduction to these processes is given next.

Vat Photopolymerisation

A 3D printer based on the Vat Photopolymerisation method has a container filled with photopolymer resin which is then hardened with a UV light source; refer to Figure-1.

Figure-1. 3D Printing via vat-photopolymerisation

The most commonly used technology in this processes is Stereolithography (SLA). This technology employs a vat of liquid ultraviolet curable photopolymer resin and an ultraviolet laser to build the object's layers one at a time. For each layer, the laser beam traces a cross-section of the part pattern on the surface of the liquid resin. Exposure to the ultraviolet laser light cures and solidifies the pattern traced on the resin and joins it to the layer below.

After the pattern has been traced, the SLA's elevator platform descends by a distance equal to the thickness of a single layer, typically 0.05 mm to 0.15 mm (0.002" to 0.006"). Then, a resin-filled blade sweeps across the cross section of the part, re-coating it with fresh material. On this new liquid surface, the subsequent layer pattern is traced, joining the previous layer. The complete three dimensional object is formed by this project. Stereolithography requires the use of supporting structures which serve to attach the part to the elevator platform and to hold the object because it floats in the basin filled with liquid resin. These are removed manually after the object is finished.

Material Jetting

In this process, material is applied in droplets through a small diameter nozzle, similar to the way a common inkjet paper printer works, but it is applied layer-by-layer to a build platform making a 3D object and then hardened by UV light; refer to Figure-2.

Figure-2. 3D Printing via Material-Jetting

Binder Jetting

With binder jetting two materials are used: powder base material and a liquid binder. In the build chamber, powder is spread in equal layers and binder is applied through jet nozzles that "glue" the powder particles in the shape of a programmed 3D object; refer to Figure-3. The finished object is "glued together" by binder remains in the container with the powder base material. After the print is finished, the remaining powder is cleaned off and used for 3D printing the next object.

Figure-3. 3D Printing via binder-jetting

Material Extrusion

The most commonly used technology in this process is Fused deposition modeling (FDM). The FDM technology works using a plastic filament or metal wire which is unwound from a coil and supplying material to an extrusion nozzle which can turn the flow on and off. The nozzle is heated to melt the material and can be moved in both horizontal and vertical directions by a numerically controlled mechanism, directly controlled by a computer-aided manufacturing (CAM) software package; refer to Figure-4. The object is produced by extruding melted material to form layers as the material hardens immediately after extrusion from the nozzle. This technology is most widely used with two plastic filament material types: ABS (Acrylonitrile Butadiene Styrene) and PLA (Polylactic acid) but many other materials are available ranging in properties from wood filed, conductive, flexible etc.

Figure-4. 3D Printing via Fused Deposition Modeling

In the above figure:
1 - nozzle ejecting molten material (plastic),
2 - deposited material (modelled part),
3 - controlled movable table.

Powder Bed Fusion

The most commonly used technology in this processes is Selective laser sintering (SLS). This technology uses a high power laser to fuse small particles of plastic, metal, ceramic or glass powders into a mass that has the desired three dimensional shape. The laser selectively fuses the powdered material by scanning the cross-sections (or layers) generated by the 3D modeling program on the surface of a powder bed; refer to Figure-5. After each cross-section is scanned, the powder bed is lowered by one layer thickness. Then a new layer of material is applied on top and the process is repeated until the object is completed.

All untouched powder remains as it is and becomes a support structure for the object. Therefore there is no need for any support structure which is an advantage over SLS and SLA. All unused powder can be used for the next print.

Figure-5. 3D Printing via Selective Laser Sintering

Sheet Lamination

Sheet lamination involves material in sheets which is bound together with external force. Sheets can be metal, paper or a form of polymer. Metal sheets are welded together by ultrasonic welding in layers and then CNC milled into a proper shape; refer to Figure-6. Paper sheets can be used also, but they are glued by adhesive glue and cut in shape by precise blades.

Directed Energy Deposition

This process is mostly used in the high-tech metal industry and in rapid manufacturing applications. The 3D printing apparatus is usually attached to a multi-axis robotic arm and consists of a nozzle that deposits metal powder or wire on a surface and an energy source (laser, electron beam or plasma arc) that melts it, forming a solid object; refer to Figure-7.

Figure-6. 3D Printing via Ultrasonic Sheet Lamination.jpg

Figure-7. 3D Printing via Direct Energy Deposition

Now, you know about different 3D Printing techniques so it is clear from different processes that the main work of 3D printer is to solidify material at your will, through different techniques. Now, we will learn about the points to be taken care of while preparing model for 3D printing.

Part Preparation for 3D Printing

Various important points to remember while preparing part for 3D printing are given next.

- You should avoid holes in the areas where model is not supported. Holes can cause material to flow out in various 3D printing processes.
- Use the Solid models. It does not mean that you cannot use surface modeling tools but you should avoid surfaces in your model. Thicken your surfaces after performing modeling operations.
- Make sure that you have not left any unwanted piece inside the model enclosure after performing boolean operations.
- Shell your model after creating it. You can 3D print solid model but if a hollow box can do your work then why to waste material on solid cube. Less material means cost efficient.
- If you want to write text on your model then check the specification of your printer for possible font range.
- The color and textures you apply on model in SolidWorks will not be exported for 3D printing so do not waste your time on them.
- 3D printing is closer to mesh modeling than solid modeling. So, check your mesh model by exporting solid model in .stl file.
- If your file has quite a bit of text and multiple emboss/engrave features, try exporting the file as a vector. Vector files are more appropriate for extremely complex files. Try .iges or .step.
- Orientation is particularly important when it comes to 3D printing in order to determine the interior and exterior of an object. Orient your part the same way as your want it in 3D Printing.
- Don't make a multi-body model for 3D printing. Your printer may die thinking what to do when two separate bodies overlap each other!!
- SolidWorks do not has capability to edit mesh model so you should prepare your model carefully to get desired mesh model.

We know the basic guidelines for preparing 3D model so it is time to export or print model on 3D Printer.

PRINT3D

The **Print3D** tool is used to print the model in drawing area on the 3D Printer. The procedure to use this tool is given next.

- After preparing model, click on the **Print3D** tool from the **File** menu. The **Print3D PropertyManager** will be displayed; refer to Figure-8. Also, you will be asked to select a flat face of the model to place it on machine bed. Note that you should have Windows 8.1 or higher installed in system to use **Print3D PropertyManager.**
- Select the desired face of the model to place it on the machine bed. The model will be displayed in a transparent box. This transparent box is the limit of your machine to 3D print; refer to Figure-9.
- Select the desired printer from the drop-down in the **Printer** rollout of the **PropertyManager** if other printer is selected.

Figure-8. Print3D PropertyManager

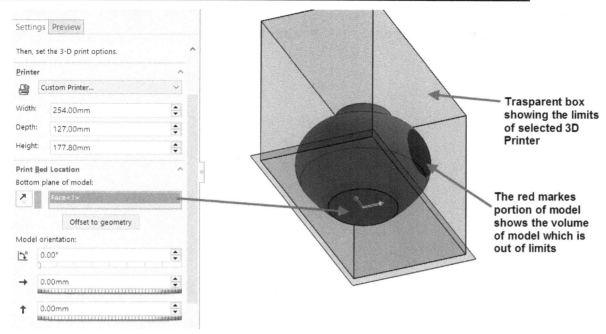

Figure-9. Preview of 3D Printer

- Set the orientation and location of part on the printer bed using the options in the **Print Bed Location** rollout of the **PropertyManager.**
- Scale your model to fit properly in printer volume by using the **Scale** edit box or **Scale to Fit** button in the **Scale** rollout.
- Set the quality of printing in the Job quality drop-down of Options rollout. Select the desired percentage of material to be filled inside the object using the **Infill percentage** drop-down.
- Select the **Include supports** check box to automatically add support in the part during 3D Printing. Support is applied where material can fall down during 3D Printing.
- Select the **Include raft** check box to include raft in the 3D Printed model. This raft of disposable material can be drained after job of printing is complete.
- Click on the **Printer Properties** button to set properties of 3D Printer.
- To save the current model in 3D Printing format file, expand the **Save To File** rollout at the bottom of **PropertyManager.** The rollout will be displayed as shown in Figure-10. Select the desired format from the drop-down and click on the **Save File** button. The **Save As** dialog box will be displayed. Save the file at desired location.

Figure-10. Save To File rollout

- To run analysis on the object to be printed, click on the **Preview** tab in the **PropertyManager**. The **PropertyManager** will be displayed as shown in Figure-11.

Figure-11. Preview tab of Print3D PropertyManager

- Set the desired parameters in the **PropertyManager** and then click on the **Calculate** button. By default, the areas which need support are displayed in green color. You can change the color by using **Edit Color** buttons. Select the **Show as transparent** check box in the **Build Analysis** rollout of **PropertyManager** to check the inside of part.
- Click on the **OK** button from the **PropertyManager** to start printing.

MODEL BASED DEFINITION (MBD)

SOLIDWORKS MBD is an integrated drawingless manufacturing solution for SOLIDWORKS. It helps to define, organize, and publish 3D Product Manufacturing Information (PMI) including 3D model data in industry standard file formats, such as eDrawings, 3D PDF and STP242. Unlike traditional 2D drawings, **SOLIDWORKS MBD** guides the manufacturing process directly in 3D, which helps streamline production, cut cycle time, reduce errors, and support industry standards such as Military-Standard-31000A, ASME Y14.41, ISO 16792, and DIN ISO 16792. The purpose of using **SOLIDWORKS MBD** is to reduce time of creating 2D drawings and hence reducing engineering cost. Using MBD, you can display the annotations in the model itself while designing. If you are going to create 2D drawings after creating model then using MBD will not be much beneficial to you. The tools for MBD are available in the **SOLIDWORKS MBD CommandManager** of the **Ribbon**; refer to Figure-12.

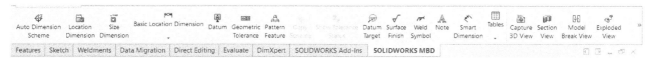

Figure-12. SOLIDWORKS MBD CommandManager

Various tools in this CommandManager are discussed next.

Auto Dimension Scheme

The **Auto Dimension Scheme** tool is used to display all the dimensions you have earlier applied to the model during sketching and other operations. The procedure to use this tool is given next. (Note: we have used the model of **Chapter 09** in this book as example for this tool.)

Figure-13. Auto Dimension Scheme PropertyManager

- Make sure you have a model in the drawing area with dimensions applied to it in sketching and other operations. Click on the **Auto Dimension Scheme** tool from the **SOLIDWORKS MBD CommandManager** in the **Ribbon**. The **Auto Dimension Scheme PropertyManager** will be displayed as shown in Figure-13.

- From the **Part type** area of **Settings** rollout, select the **Prismatic** radio button if you want the dimensions and geometric tolerances to be displayed in the form of callouts; refer to Figure-14. Select the **Turned** radio button if you want the dimensions and geometric tolerances to displayed by standard dimensioning scheme; refer to Figure-14.

- From the **Tolerance type** area of the **Settings** rollout, select the **Plus and Minus** radio button to display the dimensions with plus-minus tolerances; refer to Figure-14. Select the **Geometric** radio button to display the tolerances in geometric tolerance form; refer to Figure-15.

Prismatic Part Type selected   **Turned Part Type selected**

Figure-14. Dimensioning by different part types

Figure-15. Dimensioning by geometric tolerance with Turned part type

- Select the **Linear** radio button from **Datum Selection** area to display the linear pattern in dimensioning. In this way, polar patterns will be dimensioned as linear instances. Select the **Polar** radio button to display polar pattern in dimensioning; refer to Figure-16. After selecting the **Polar** radio button, set the minimum number of instances to be taken as pattern in the spinner next to radio button.

Linear Pattern   Polar Pattern

Figure-16. Pattern options for Dimensioning

- Click in the **Primary Datum** selection box of **Datum Selection** rollout and select the reference face/plane/axis to be the primary datum for geometric tolerance. Similarly, select the Secondary and Tertiary datums; refer to Figure-17.

Figure-17. Datum reference selection

- Note that on selecting reference, a small toolbar is displayed along the cursor. Select the appropriate button from toolbar while selecting datum reference.
- Set the scope for dimensioning and select check boxes to include features for dimensioning from the **Feature Filters** rollout.
- Click on the **OK** button from the **PropertyManager** to apply dimensioning. Note that the dimension history will be displayed in the **DimXpertManager** at the right in the application window.

Location Dimension

The **Location Dimension** tool is used to create location dimension for various features in the model. The procedure to use this tool is given next.

- Click on the **Location Dimension** tool from the **SOLIDWORKS MBD CommandManager**. You will be asked to select the face(s) to apply dimension.
- Select the faces/axes to be dimensioned. The dimension will get attached to cursor; refer to Figure-18.

Figure-18. Location Dimensioning

- Click at the desired location to place the dimension. The dimension will be created and **DimXpert PropertyManager** will be displayed; refer to Figure-19.
- Set the desired parameters and click on the **OK** button. Most of the options in this **PropertyManager** are same as discussed for **Dimension PropertyManager** in previous chapters.

Size Dimension

The **Size Dimension** tool is used to apply size dimensions to various features in the model. The procedure to use Size Dimension tool is given next.

- Click on the **Size Dimension** tool from the **SOLIDWORKS MBD CommandManager**. You will be asked to select the faces/feature to be dimensioned.

Figure-19. DimXpert Property-Manager

- Select the feature to be dimensioned. The dimension will get attached to cursor and a toolbar will be displayed; refer to Figure-20.

Pattern button selected
from toolbar

Cylinder/hole button
selected from toolbar

Figure-20. Creating size dimension

- Select the desired button from the toolbar to modify the dimension accordingly.
- Click in the drawing area to place the dimension. The **DimXpert PropertyManager** will be displayed. Set the desired options and then click on the **OK** button.

Basic Location Dimension/Basic Size Dimension

The **Basic Location Dimension** and **Basic Size Dimension** tools work in the same way as **Location Dimension** and **Size Dimension** tools respectively. Basic dimensions by nature are a theoretically exact value; however, the feature(s) of a part they define as ideal or exact do need to have tolerances to permit acceptable levels of imperfection during manufacturing.

When a feature is defined with basic dimensions, the tolerance for that feature must be expressed through a geometric tolerance. Most often, the geometric tolerance is indicated directly to the feature or feature of size on the face of the drawing; however, some companies include a general geometrical tolerance (such as a position tolerance or profile of a surface tolerance) in the drawing's general notes. This can be an effective tool when the note is carefully written.

Drawings based on ISO standards frequently use a class of general geometrical tolerances standardized in ISO 2768-2 :1989. The three classes are identified through the use of the upper case letters H, K, or L after the ISO 2768 indication on the drawing. You would need a copy of the standard to interpret the amount of tolerance available.

Datum

The **Datum** tool is used to apply datum reference symbol to selected face/axis. The procedure to use this tool is given next.

- Click on the **Datum** tool from the **SOLIDWORKS MBD CommandManager**. The Datum symbol will get attached to cursor and the **Datum Feature PropertyManager** will be displayed; refer to Figure-21.

Figure-21. Datum Feature PropertyManager with Datum symbol attached to cursor

- The options in the **PropertyManager** are same as discussed in Chapter-09. Click on the face/axis of model and click at desired location to place the symbol.
- Press **ESC** to exit the tool.

Geometric Tolerance

The **Geometric Tolerance** tool is used to apply geometric dimensioning and tolerance to the selected feature. The tool works in the same way as discussed in Chapter-09 of this book.

Pattern Feature

The **Pattern Feature** tool is used to form a pattern group of same features at different locations. For example, you have created 5 holes on a plate at different locations without using pattern tools. Out of which two holes are in one hole group and three holes are in another hole group; refer to Figure-22. Now if you generate size dimension for holes then there will be two dimensions as shown in Figure-23 (a). If you combine all the holes in one pattern feature then the size dimension will be generated as shown in Figure-23 (b).

Figure-22. Holes created in different groups

(a) Before using Pattern Feature tool   **(b) After using Pattern Feature tool**

Figure-23. Use of Pattern Feature tool

The procedure to use this tool is given next.

- Click on the **Pattern Feature** tool from the **CommandManager**. The **DimXpert Pattern/Collection PropertyManager** will be displayed; refer to Figure-24.
- Select the **Linked Patterns** radio button to select one by one the already created patterns to combine them. Select the **Manual Patterns** radio button to select each of the feature individually. On selecting the **Manual Pattern**s radio button, the **Find all on same face** check box is displayed. Select this check box if you want to select all the features of same time on the current face. Select the **Collection** radio button to make a collection of features selected. Note that collection is just grouping of features, it does not reflect in dimensioning.
- After selecting the desired radio button, select the features from the model and click on the **OK** button.

Figure-24. DimXpert Pattern Collection PropertyManager

Show Tolerance Status

The **Show Tolerance Status** tool is used to highlight the features/ faces to which tolerances have been applied. Click on this button to toggle tolerance status. The **Green** colored features/faces show fully constrained features, the **Yellow** colored features/faces show under constrained features and **Red** colored features/faces show over constrained features; refer to Figure-25.

Figure-25. Tolerance status

Other DimXpert Tools

The other tools in the **CommandManager** viz. **Datum Target, Surface Finish, Weld Symbol, Balloon, Stacked Balloons,** and **Note** work in the same way as they do in Drawing environment. These tools have been discussed in **Chapter 09** of the book. The **Smart Dimension** tool in **SOLIDWORKS MBD CommandManager** is used to invoke other dimensioning tools of **DimXpert**.

Inserting Tables

All the tools in **Tables** drop-down of the **CommandManager** work in the same way. Here, we will discuss the method to insert Title Block Table, you can use the same method for inserting other tables. The procedure to insert Title Block Table is given next.

*   Click on the **Title Block Table** tool from the **Tables** drop-down in the **SOLIDWORKS MBD CommandManager**. The **Title Block Table PropertyManager** will be displayed; refer to Figure-26.
*   Click on the **Open template for Title Block Table** button (highlighted by cursor in Figure-26) from the **PropertyManager**. The Open dialog box will be displayed.
*   Select the desired template of title block from the dialog box and click on the **Open** button. The title block template will be added in the **Table Template** edit box of the **PropertyManager**.
*   Set the thickness of table borders and internal lines using the drop-downs in the **Border** rollout of the **PropertyManager**.

- Click on the **OK** button from the **PropertyManager**. If your part is not oriented to any standard view viz. **Front, Top** or **Right** then the **Select Annotation View** dialog box will be displayed; refer to Figure-27.

Figure-26. Title Block Table PropertyManager

Figure-27. Select Annotation View dialog box

- Click on the drop-down next to **Existing Annotation View** radio button and select the desired plane in which you want to place the table.
- Click on the **OK** button from the dialog box. The table will get attached to cursor.
- Click at the desired location to place the table. Double-click in the fields of table and enter desired text; refer to Figure-28.

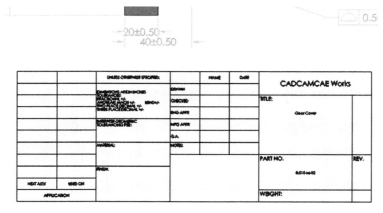

Figure-28. Title block inserted in front plane

Section View

The **Section View** tool as the name suggests, is used to section the part/assembly. Section view is sometimes very important to see the internal functioning of an assembly like in the engines. The procedure to use **Section View** tool is discussed next.

- Make sure you have a part/assembly in the drawing area. Click on the **Section View** tool from the **SOLIDWORKS MBD CommandManager**. The **Section View PropertyManager** will be displayed with preview of section of an assembly; refer to Figure-29.

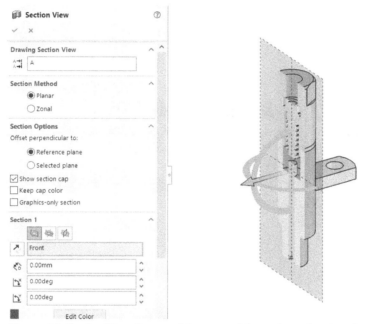

Figure-29. Section View PropertyManager with preview of sectioned assembly

- Specify the naming letter for section view in the **Drawing Section View** edit box in the **PropertyManager**.
- Select the **Show section cap** to make sure that the section view is colored and do not look like surface model. Select the **Keep cap color** check box if you want to display color on the cap applied to section view.
- Select the desired place from the **Section 1** rollout to change the plane of section. You can select the three buttons in the rollout or you can select any face from the model to set the section plane.
- Similarly, set the other options in the rollout. If you want to create section of model from two sides then select the **Section 2** check box. The **Section 2** rollout will expand. Set the desired parameters in the rollout as discussed earlier.
- Click on the **OK** button to create the section view; refer to Figure-30. Click on the **Section View** button from the **CommandManager** again to exit the section view.

Figure-30. Section view created

Model Break View

The **Model Break View** tool is used to display the model broken by specified plane-gap. The procedure to use this tool is given next.

- Click on the **Model Break View** tool from the **CommandManager**. The **Model Break View PropertyManager** will be displayed; refer to Figure-31.

Figure-31. Model Break View PropertyManager with preview of planes

- Select the desired plane from the **Alignment** rollout of the **PropertyManager** and enter the parameters like angle along X or Y of the plane, or distance between the two planes displayed in preview. Note that the area between the two planes will be cut.
- Select the **Preview** check box to see the preview while editing values.
- Click on the **Apply** button to apply current break step. If you want to create multiple breaks then click on the **Add Step** button from the **PropertyManager**; refer to Figure-32.

Figure-32. Multiple break created in model

- Click on the Styles tab in the PropertyManager and select the desired style button to check the break view style; refer to Figure-33.

Figure-33. Styles tab in PropertyManager

- Click on the **OK** button from the **PropertyManager** to create the model break view. The new view will be added in the **ConfigurationManager**.

The **Exploded View** tool works in the same way as discussed in **Chapter 7** of this book.

Capture 3D View

While going through previous chapters, you have created 2D views of model like front view, right view, top view etc. In SolidWorks, you can create 3D views to represent various aspects of model in 3D. For example, you want the save a model view with sections done on the assembly or you want to exploded view of model in 3D. These type of 3D views can be created by **Capture 3D View** tool. The procedure to use this tool is given next.

- Orient the part as you want it to be in the 3D view. Click on the **Capture 3D View** tool from the **SOLIDWORKS MBD CommandManager**. The **Capture 3D View PropertyManager** will be displayed; refer to Figure-34.

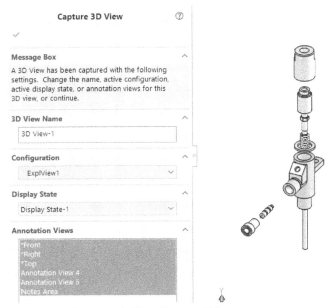

Figure-34. Capture 3D View PropertyManager

- Specify the name of view in the 3D View Name edit box of the **PropertyManager**.
- Select the views from **Annotation Views** rollout to include their annotations in the current 3D view.
- Click on the **OK** button. A new 3D view will be added to the **3D Views** tab at the bottom in the application window; refer to Figure-35.

Figure-35. 3D view added

- Change the orientation or make a section view and then click on the **Capture 3D View** tool again to create another 3D view.

Dynamic Annotation Views

The **Dynamic Annotation Views** tool is used to make the annotations dynamic with respect to selected view. For example, if you have applied annotations to the model in front and right view, and you activate the **Dynamic Annotation Views** tool then you will not see the annotation of right view when model is in front view. Click on the **Dynamic Annotation**

Views tool from the **SOLIDWORKS MBD CommandManager** to toggle and rotate the model in various views to check annotations relevant to current view only; refer to Figure-36.

Before activating Dynamic Annotation Views tool   **After activating Dynamic Annotation Views tool**

Figure-36. Result of selecting Dynamic Annotation Views tool

Publish to 3D PDF

The **Publish to 3D PDF** tool is used to create 3D pdf file from the model which can be distributed to clients who are not having CAD software in their system. The procedure to use this tool is given next.

- Click on the **Publish to 3D PDF** tool from the **SOLIDWORKS MBD CommandManager**. The **Publish to 3D PDF PropertyManager** will be displayed; refer to Figure-37.
- Click on the **Theme** drop-down in the **Theme** rollout and select the desired theme for creating 3D PDF.
- Select the desired button from **Include Primary Views** rollout to make it primary view in the PDF.
- Select the other views to be included in pdf from More views box of Include Primary Views rollout while holding the CTRL key.
- Select the desired accuracy level from the **Accuracy** rollout of the **PropertyManager**. Higher accuracy means bigger size of pdf file generated.
- Select the **Create and attach STEP-242** check box to attach the STEP-242 format file with the pdf. The STEP-242 enables to export the tolerance and other manufacturing data with model.
- Click on the **Next** button at the top in the **Publish to 3D PDF PropertyManager**. The **PropertyManager** will be displayed as shown in Figure-38.
- Click on the **Edit** button from the **PropertyManager**. The **Custom Property Text** dialog box will be displayed; refer to Figure-39.

Figure-37. Publish to 3D PDF PropertyManager

Figure-38. Output Text Fields page in PropertyManager

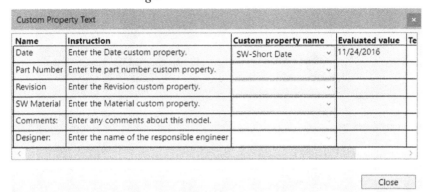

Figure-39. Custom Property Text dialog box

- Perform editing of fields which are edit-able and click on the **Close** button.
- Click on the **OK** button from the **PropertyManager**. The **Save As** dialog box will be displayed; refer to Figure-40.

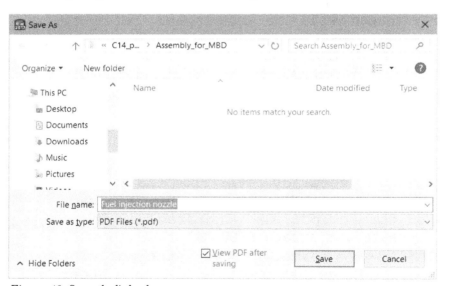

Figure-40. Save As dialog box

- Select the **View PDF after saving** check box to open the pdf after creation.
- Specify the desired name of file, move to the desired directory, and click on the **Save** button to save the file. The 3D PDF file will be created; refer to Figure-41.

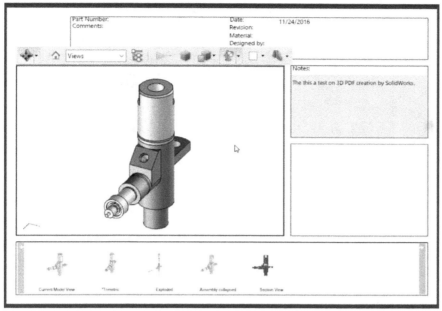

Figure-41. Sample 3D PDF created

- You can perform various actions in the 3D PDF like zoom, pan, rotate, and spin the model. You can also measure various parameters in the PDF itself using the tools in PDF reader.

3D PDF Template Editor

The **3D PDF Template Editor** tool is used to edit/create the templates using which the 3D PDFs are created. The procedure to use 3**D PDF Template Editor** tool is given next.

- Click on the **3D PDF Template Editor** tool from the **CommandManager**. The **Template Editor** window will be displayed; refer to Figure-42.

Figure-42. Template Editor window

- Click at the item you want to change. The respective editing box will be displayed; refer to Figure-43.

Figure-43. Editing box for fields in 3D PDF template

- Set the desired parameters and click on the **Save/Save As** button to save the template.

Now, you can use the newly created template to generate 3D PDFs. Note that you need to start SolidWorks with Administrator rights to save the template file in system directories.

In the same way, you can publish eDrawings file and STEP 242 file.

3D PMI Compare

The **3D PMI Compare** tool is used to compare the 3D product and manufacturing information (PMI) of two documents. It compares DimXpert annotations, reference dimensions, and other annotations between two part documents. The procedure to use this tool is given next.

- Click on the **3D PMI Compare** tool from the **CommandManager**. The **3D PMI Compare** task pane will be displayed on the right in the application window; refer to Figure-44.
- Click on the browse button next to **Reference document** drop-down in the task pane and open the model you want to compare.
- In the same way, select the document for **Modified document** drop-down. The **Run Comparison** button will become active.
- Click on the **Run Comparison** button. The PMI compare list will be displayed in the task pane; refer to Figure-45.

Figure-44. 3D PMI Compare task pane

Figure-45. PMI Compare list

- Save the report by using Save Report button in the task pane, if required.
- Click on the **Close** button in the task pane to exit.

FOR STUDENT NOTES

Chapter 16

Introduction to SolidCAM

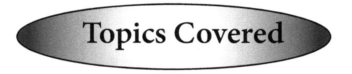
Topics Covered

The major topics covered in this chapter are:

- *Defining Machine*
- *Setting Stocks and Parts*
- *Extracting Machinable Features*
- *Performing Operations*
- *Generating CNC Program*

INTRODUCTION

SolidWorks CAM is a new front introduced in SolidWorks 2018. SolidWorks CAM is based on CAMWorks software. This software is available as Add-In for SolidWorks so you can activate or deactivate it as discussed earlier. Being an Add-In it provides a well integrated mode of operation within SolidWorks. The software recognizes geometric tolerance and other dimensional features of the model created in SolidWorks. By default the **SolidWorks CAM** Add-In is active in SolidWorks 2018 and the tools related to CAM are available in **SOLIDWORKS CAM CommandManager**; refer to Figure-1.

Figure-1. SOLIDWORKS CAM CommandManager

Note that various trees are available for SolidWorks CAM in the left area of the application window; refer to Figure-2.

SOLIDWORKS CAM
Feature Tree

SOLIDWORKS CAM
Operation Tree

SOLIDWORKS CAM
Tool Tree

Figure-2. SOLIDWORKS CAM Trees

Various tools and procedures of SolidWorks CAM are discussed next.

DEFINING MACHINE

The **Define Machine** tool is used to define the type of machine and other related parameters for creating machine setup as per the job. The procedure to define machine is given next.

- After opening the part file, click on the **Define Machine** tool from the **SOLIDWORKS CAM CommandManager**. The **Machine** dialog box will be displayed; refer to Figure-3.

Figure-3. Machine dialog box

- Select the desired machine from the list of machine. Note that the machines will be active only based on your part. Set the parameters related to machine.

Setting Tool Crib

- Click on the **Tool Crib** tab from the dialog box. The dialog box will be displayed as shown in Figure-4.
- There are 12 tools by default in the tool crib. You can add or remove the tools based on your machine's tool cassette.

Adding Tools

- To add a tool, click on the **Add Tool** button from the dialog box. The **Tool Select Filter** dialog box will be displayed; refer to Figure-5.

Figure-4. Tool Crib tab in Machine dialog box

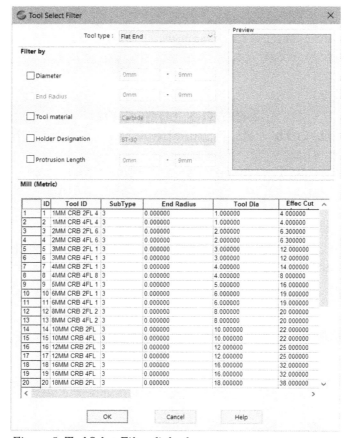

Figure-5. Tool Select Filter dialog box

- Click in the **Tool type** drop-down and select the desired type of tool to be added in the crib. Like if you want to use Ball Nose cutting tool then select the Ball Nose tool from the drop-down.
- Set the desired filters for the tool and select the desired tool to be added.
- Click on the **OK** button from the dialog box. The tool will be added in the list and the preview of tool will be displayed; refer to Figure-6.

Figure-6. Preview of the tools

Editing Tool

- Select the tool you want to edit from the list and click on the **Edit Tool** button. The **Edit Tool Parameters** dialog box will be displayed; refer to Figure-7.
- Set the desired parameters in the dialog box. Note that you can set the tool holder and tool station as well for your tool.
- After setting the tool, click on the **OK** button.

Figure-7. Edit Tool Parameters dialog box

Removing Tool from the Crib

- Select the tool that you want to remove from crib and click on the **Remove Tool** button from the dialog box. The tool will be removed.

Updating tool in Database

Based on the parameters defined by you in the dialog box, you can update the definitions of the tools in the central database of SolidWorks. To do so, after editing tools; click on the **Update Tool** button from the dialog box.

Saving Tool Crib

After creating the desired tool crib, you can save it for later use. The procedure is given next.

- Click on the **Save Tool Crib** button from the dialog box. The **Save to Database** dialog box will be displayed; refer to Figure-8.
- Set the desired name and parameters and then click on the **Save** button.

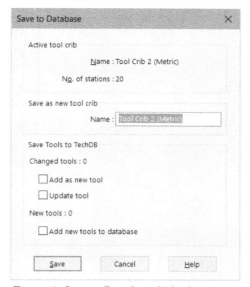

Figure-8. Save to Database dialog box

Creating and Saving New Tool in Library

- Click on the **New Tool** button from the **Tool library** area of the dialog box. The **New Tool** dialog box will be displayed similar to **Edit Tool Parameters** dialog box; refer to Figure-9.

Figure-9. New Tool dialog box

- As the parameters as discussed earlier and click on the **OK** button to create the tool. The tool will be created and added to the library as well as tool crib.
- If a tool is not added in the library, then select it from the crib table and click on the **Save Tool** button from the **Tool library** area.

Setting Postprocessor Parameters

The post processor is used to translate the codes generated by CAM program to machine readable codes based on machine controller. The options for post processor are available in the **Post Processor** tab of the **Machine** dialog box; refer to Figure-10.

Figure-10. Post Processor tab of Machine dialog box

- Select the desired post processor from the list and click on the Select button.
- If you have an updated post processor then click on the **Browse** button. The **Open** dialog box will be displayed.
- Select the desired post processor and click on the **Open** button.

Posting Options

The options related to posting coolant offset etc. are available in the **Post** tab; refer to Figure-11.

Figure-11. Posting tab of Machine dialog box

- Set the desired parameters and click on the **OK** button to apply the parameters.

You can set the other parameters of the dialog box as discussed earlier.

SETTING COORDINATE SYSTEM

The coordinate system is an important part of CNC machines programming. All the coordinates for machine programming are referenced to the coordinate system selected. The procedure to set coordinate system is given next.

- Click on the **Coordinate System** tool from the **SOLIDWORKS CAM CommandManager**. The **Fixture Coordinate System PropertyManager** will be displayed; refer to Figure-12. Also, you will be asked to select a point to define the coordinate system.

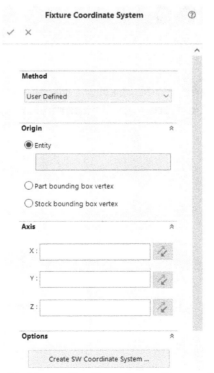

Figure-12. Fixture Coordinate System PropertyManager

- Select a point on the model that you want to be coordinate system. To align the coordinate system at desired orientation, click in the respective direction selection box of the **Axis** rollout and select the desired reference.
- If you want to use any vertex of the part bounding box then select the **Part bounding box vertex** radio button. Vertices will be displayed; refer to Figure-13.

Figure-13. Bounding box vertices

- Select the desired vertex to place coordinate system.
- If you want to use a vertex of bounding box of stock as coordinate system reference then select **Stock bounding box vertex** radio button.
- After selecting the point and setting the direction references, click on the **OK** button from the **PropertyManager** to create the coordinate system.

DEFINING OR EDITING STOCK

Stock is the pile of material to be removed after machining to produce the part. The procedure to define/edit the stock is given next.

- Click on the **Stock Manager** tool from the **SOLIDWORKS CAM CommandManager** in the **Ribbon**. The **Stock Manager PropertyManager** will be displayed; refer to Figure-14.

Figure-14. Stock Manager Property-Manager

- Select the desired material from the drop-down.
- There are four options to set the stock type in the **Stock Type** rollout; **Bounding Box**, **Extruded Sketch**, **STL File**, and **Part File**. Select the **Bounding Box** button if you want to create a stock bounding the whole workpiece.
- Select the **Extruded Sketch** button if you want to create stock by extruding the selected sketch; refer to Figure-15.
- Select the **STL File** button and click on the **Browse** button in the **STL File** rollout. The **Open** dialog box will be displayed; refer to Figure-16. Select the desired file and click on the **Open** button.

Figure-15. Extruded sketch stock

Figure-16. Open dialog box

- Click on the **Part File** button from the rollout if you want to use a Solidworks file as stock. After selecting the button, click on the **Browse** button. The **Open** dialog box will be displayed. Select the desired file and click on the **Open** button from the dialog box. The part will be placed at the origin.
- After creating the stock and specifying the desired parameters, click on the **OK** button. The stock will be created.

SETTING MILLING OPERATION PARAMETERS

The milling operations require a few parameters to be set before performing cutting operations. The procedure to set the parameters are given next.

- Click on the **Mill Setup** tool from the **Setup** drop-down in the **SOLIDWORKS CAM CommandManager**. The **Mill Setup PropertyManager** will be displayed; refer to Figure-17.

Figure-17. Mill Setup Property-Manager

- Click in the selection box of **Entity** rollout and select the face/plane perpendicular to which the tool will be aligned while cutting. In simple words, select the desired face/plane to define cutting direction.
- Select the **Associate** check box if you want the setup to be modified with the model if any changes are made to it.
- Select the desired feature from the **Features** rollout. Like, select the **Face** check box if you want SolidWorks to automatically identify the face features for machining. Select the **Perimeter** check box if you boss/open pocket features to be identified. Select the **Multi surface feature** check box if you want to identify multi surface features.
- Click on the **OK** button from the **PropertyManager** to apply the settings.

EXTRACTING MACHINABLE FEATURES

The **Extract Machinable Features** tool is used to extract all the machinable features based on the cutting strategies in the database. The procedure to use this tool is given next.

- Click on the **Extract Machinable Features** tool from the **Ribbon**. The features that are machinable will be created; refer to Figure-18.

Figure-18. Automatically identified feature

GENERATING OPERATION PLAN

The **Generate Operation Plan** tool is used to create cutting strategies based on identified features. To generate cutting strategies, click on the **Generate Operation Plan** tool from the **SOLIDWORKS CAM CommandManager** in the **Ribbon**. The material cutting strategies will be created; refer to Figure-19.

Figure-19. Material cutting strategies created

GENERATING TOOLPATHS

The **Generate Toolpath** tool is used to create toolpaths based on earlier generated operation plan. Click on the tool and toolpaths will be generated based on tools and parameters earlier specified.

SIMULATING TOOLPATH

To simulation the toolpaths, click on the **Simulate Toolpath** tool from the **Ribbon**. If there is no toolpath selected then all the toolpaths will be selected automatically for simulation. On selecting this tool, the **Toolpath Simulation PropertyManager** will be displayed; refer to Figure-20.

Figure-20. Toolpath Simulation PropertyManager

• Click on the **Play** button to simulate toolpath.

VISUALIZING TOOLPATHS

The **Step Thru Toolpath** tool is used to display toolpaths. On clicking the tool, the **Step Through Toolpath PropertyManager** will be displayed with the preview of toolpaths; refer to Figure-21.

Figure-21. Step Through Toolpath PropertyManager

Click on the **Play** button from the **PropertyManager** to check each of the toolpath simulation.

SAVING TOOLPATHS

The **Save CL File** tool is used to save the toolpaths in CL file format. The procedure is given next.

- Click on the **Save CL File** tool from the **SOLIDWORKS CAM CommandManager** in the **Ribbon**. The **Save As** dialog box will be displayed.
- Specify the desired name and location. Click on the **Save** button to save the file.

POST PROCESSING TOOLPATHS

Post processing is the process of generating G codes from the toolpaths. The procedure to do so is given next.

- Click on the **Post Process** tool from the **Ribbon**. The **Post Output File** dialog box will be displayed; refer to Figure-22.

Figure-22. Post Output File dialog box

- Specify the desired name and click on the **Save** button. The **Post Process PropertyManager** will be displayed; refer to Figure-23.
- Select the **Open G-Code file in** check box if you want to edit the G-codes after creating them.
- Click on the **Play** button from the **PropertyManager** to start creating G-codes. The codes will be displayed in the **NC Codes** box.

Figure-23. Post Process PropertyManager

- Click on the **OK** button from the **PropertyManager**. The **SOLIDWORKS CAM NC Editor** will be displayed if you have selected the **Open G-Code file in** check box; refer to Figure-24.
- Modify the file as required and save it. Close the application.

Figure-24. SOLIDWORKS CAM NC Editor window

MANUALLY CREATING MILL OPERATIONS

In previous sections, you have learned to automatically create milling operations. Now, we will discuss the procedure of creating milling operations manually. We will work with 2.5 Axis Mill Operations and you can apply the procedure to other tools.

- Click on the **2.5 Axis Mill Operations** tool from the **SOLIDWORKS CAM CommandManager** in the **Ribbon**. The **New Operation PropertyManager** will be displayed; refer to Figure-25.

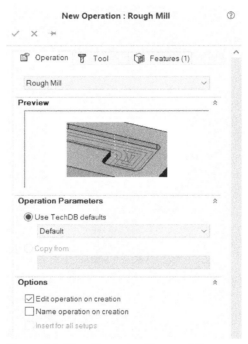

Figure-25. New Operation PropertyManager

- Select the desired operation type from the drop-down at the top in the **PropertyManager**. Select the **Rough Mill** option to perform rough milling operation for large stock. Select the **Contour Mill** option to remove material around the edges of boss/pocket features. Select the **Face Mill** option to remove material from the top face of a part. Select the **Thread Mill** option to machine threads on boss or hole features. We are using the **Rough Mill** option in our case.
- Click on the **Tool** tab in the **PropertyManager**. The options will be displayed as shown in Figure-26.

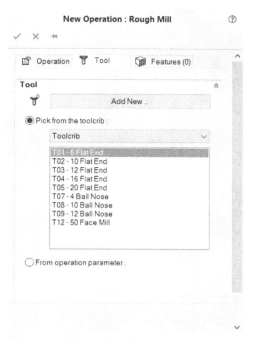

Figure-26. Tools tab in New Operation PropertyManager

- Select the desired tool from the list.
- Click on the **Features** tab in the **PropertyManager**. The options will be displayed as shown in Figure-27.

Figure-27. Features tab in New Operation PropertyManager

- Select the desired feature from the **Pick from the available** box in the **Features for the Rough Mill** rollout if you have earlier applied **Extract Machinable Features** tool. If you have not used the tool earlier the you can create features now also.
- Click on the **Create Features** button from the **PropertyManager** and then click on the **2.5 Axis Feature** option. The **2.5 Axis Feature: Select Entities PropertyManager** will be displayed; refer to Figure-28.

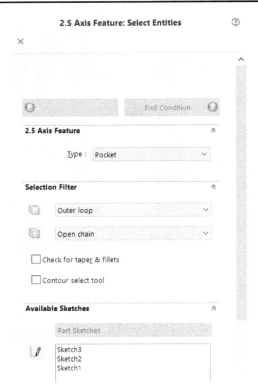

Figure-28. 2.5 Axis Feature: Select Entities PropertyManager

- Select the desired feature type from the **Type** drop-down of **2.5 Axis Feature** rollout. Set the selection filters as required from the **Selection Filter** rollout.
- Click on the face(s) to be used to create the feature; refer to Figure-29.

Figure-29. Face selected for creating mill features

- Click on the **End Condition** button at the top in the **PropertyManager**. The options will be displayed as shown in Figure-30.

Figure-30. End condition options

- Set the desired parameters and click on the **OK** button. The feature will be created.
- After creating the feature and selecting it, click on the **OK** button from the **New Operation PropertyManager**. The cutting operation will be created and the **Operation Parameters** dialog box will be displayed; refer to Figure-31.
- Set the parameters as required and click on the **OK** button to create the operation.
- You can now generate the toolpath as discussed earlier.

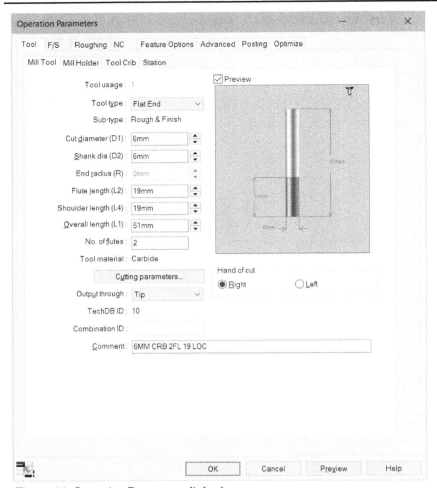

Figure-31. Operation Parameters dialog box

Till this point, we have learned about the basics of SolidWorks. Since the book is not dedicated to SolidWorks CAM so we have given just introduction here.

Index

Ethics of an Engineer

- Engineers shall hold paramount the safety, health and welfare of the public and shall strive to comply with the principles of sustainable development in the performance of their professional duties.

- Engineers shall perform services only in areas of their competence.

- Engineers shall issue public statements only in an objective and truthful manner.

- Engineers shall act in professional manners for each employer or client as faithful agents or trustees, and shall avoid conflicts of interest.

- Engineers shall build their professional reputation on the merit of their services and shall not compete unfairly with others.

- Engineers shall act in such a manner as to uphold and enhance the honor, integrity, and dignity of the engineering profession and shall act with zero-tolerance for bribery, fraud, and corruption.

- Engineers shall continue their professional development throughout their careers, and shall provide opportunities for the professional development of those engineers under their supervision.

OTHER BOOKS BY CADCAMCAE WORKS

Autodesk Inventor 2018 Black Book

Autodesk Fusion 360 Black Book

AutoCAD Electrical 2015 Black Book
AutoCAD Electrical 2016 Black Book
AutoCAD Electrical 2017 Black Book
AutoCAD Electrical 2018 Black Book

SolidWorks 2014 Black Book
SolidWorks 2015 Black Book
SolidWorks 2016 Black Book
SolidWorks 2017 Black Book

SolidWorks Simulation 2015 Black Book
SolidWorks Simulation 2016 Black Book
SolidWorks Simulation 2017 Black Book

SolidWorks Electrical 2015 Black Book
SolidWorks Electrical 2016 Black Book
SolidWorks Electrical 2017 Black Book

Mastercam X7 for SolidWorks 2014 Black Book
Mastercam 2017 for SolidWorks Black Book

Creo Parametric 3.0 Black Book
Creo Parametric 4.0 Black Book

Creo Manufacturing 4.0 Black Book

www.ingramcontent.com/pod-product-compliance
Lightning Source LLC
Chambersburg PA
CBHW080546060326
40689CB00021B/4766